Student Solutions Manual

for use with

Beiser's
Concepts of
Modern Physics

Sixth Edition

Craig B. Watkins
Massachusetts Institute of Technology

Boston   Burr Ridge, IL   Dubuque, IA   Madison, WI   New York   San Francisco   St. Louis
Bangkok   Bogotá   Caracas   Lisbon   London   Madrid
Mexico City   Milan New Delhi   Seoul   Singapore   Sydney   Taipei   Toronto

McGraw-Hill Higher Education
A Division of The McGraw-Hill Companies

Student Solutions Manual for use with
Beiser's CONCEPTS OF MODERN PHYSICS, SIXTH EDITION
CRAIG B. WATKINS

Published by McGraw-Hill Higher Education, an imprint of The McGraw-Hill Companies, Inc., 1221 Avenue of the Americas, New York, NY 10020. Copyright © The McGraw-Hill Companies, Inc., 2003, 1995. All rights reserved.

No part of this publication may be reproduced or distributed in any form or by any means, or stored in a database or retrieval system, without the prior written consent of The McGraw-Hill Companies, Inc., including, but not limited to, network or other electronic storage or transmission, or broadcast for distance learning.

This book is printed on acid-free paper.

6 7 8 9 0 QSR QSR 0 9 8 7

ISBN-13: 978-0-07-249971-1
ISBN-10: 0-07-249971-0

ISBN 0-07-249971-0

www.mhhe.com

Preface to the Student's Solutions Manual

For the sixth edition of **Concepts of Modern Physics**, most problem solutions have not changed substantially. Several considerations that have been made reflect a further trend towards greater use of more sophisticated calculational techniques, from more elaborate handheld devices to symbolic manipulation and graphing software. None of the problems or exercises in the text require powerful tools, but it is expected that those with access to these tools will be inclined to use them. When such methods might be advantageous for specific problems, the solution to those problems try to identify the methods.

With few exceptions, the values of physical constants used are from the front endpapers or the Appendix of atomic masses. In doing the calculations for this solutions manual, extra significant figures were retained in intermediate calculations, and the answers rounded to the appropriate number of significant figures. For several of the problems, additional features have been used as alternatives.

- **Reference to Other Sources** has been kept to a minimum; the text itself contains all of the information needed. If values of physical constants are desired to better accuracy, a suggested source is the **Particle Data Group** tables of **Constants, Units, Atomic and Nuclear Properties**, available at

  http://pdg.lbl.gov/2001/contents_sports.html

- **Plotting and Analysis of Data** problems were done using a spreadsheet program, primarily for the convenience of obtaining legible plots. The text problems that do require fits to data sets can be done to the desired precision by hand-graphing, and those with statistics capabilities on handheld calculators will be able to find least-squares best-fit linear fits to the desired precision.

- **Symbolic Manipulation and Graphics Programs** make some problems simpler, but are not required for any specific problem. For some problem solutions, examples of commands that will run on most releases of MAPLE are given.

- **Special Methods of Integration** are given for some problems, including numerical integration of functions that would otherwise rely on judicious approximation. The special analytic methods given are mainly for finding moments of distributions, using techniques familiar to physicists.

- **Numerical Approximations** for some problems requiring a high degree of precision are available in some circumstances; when these methods are included, the analytic approximations, usually involving Taylor Series, are intended as the primary method of solution.

Mistakes and transcription errors may be reduced but are unlikely to be eliminated. All problems have been rechecked from the fifth edition, and many (too many, of course) errors have been corrected. "Solutions" to the shorter odd-numbered discussion-type questions do not differ substantially from the answers at the back of the text. Any corrections or suggestions for improvements will be appreciated.

Thanks to Professor Beiser for cross-checking the rough answers, and many thanks to the folks at the **Experimntal Study Group** at **MIT** for their help, advice and patience.

Craig Watkins

MIT Experimental Study Group

January 2002

Table of Contents

Preface to the Instructor's Manual iii

Chapter 1 - Relativity 1
    Approximation Techniques 1

Chapter 2 - Particle Properties of Waves 14
    Extra Expressions for Compton Effect Problems 18

Chapter 3 - Wave Properties of Particles 25

Chapter 4 - Atomic Structure 32

Chapter 5 - Quantum Mechanics 41
    Special Integrals for Harmonic Oscillators 47
    More on Integrals for Harmonic Oscillators 52
    Integrals for a Particle in a Box 54

Chapter 6 - Quantum Theory of the Hydrogen Atom 55
    Integrals for Hydrogen Wave Functions 61

Chapter 7 - Many-Electron Atoms 63

Chapter 8 - Molecules 67

Chapter 9 - Statistical Mechanics 71

Chapter 10 - The Solid State 81

Chapter 11 - Nuclear Structure 86

Chapter 12 - Nuclear Transformations 92

Chapter 13 - Elementary Particles 105

Chapter 1 - Relativity

Problems that involve relativistic effects at speeds much smaller than the speed of light, or the equivalence of special relativity and Newtonian mechanics at low speeds, often require finding differences such as

$$1 - \sqrt{1 - \frac{v^2}{c^2}} \quad \text{or} \quad \frac{1}{\sqrt{1 - \frac{v^2}{c^2}}} - 1$$

when $v \ll c$.

These are both differences between quantities that are equal to 1 in the limit as $v \to 0$, but as the quantities are not the same for $v \neq 0$, we are interested in how the differences depend on v (more specifically, the ratio v/c) in the limit $v \ll c$.

There are many ways to find the functional form of these differences; four familiar methods are explained here.

I - Binomial Theorem for Non-integral Exponents

This is the method used in Section 1.8.

A familiar form of the binomial theorem is

$$(1+x)^\alpha = 1 + \alpha x + \frac{\alpha(\alpha-1)}{2} x^2 + \frac{\alpha(\alpha-1)(\alpha-2)}{2 \cdot 3} x^3 + \cdots.$$

If α is a nonnegative integer, the coefficients of the powers of x are the usual binomial coefficients, and the series truncates. However, if $|x| < 1$, the series will converge for other values of α, particularly negative integers or fractions. Specifially, if $\alpha = -\frac{1}{2}$,

$$(1+x)^{-1/2} = 1 + \left(-\frac{1}{2}\right)x + \frac{(-1/2)(-3/2)}{2} x^2 + \frac{(-1/2)(-3/2)(-5/2)}{6} x^3 + \cdots$$
$$= 1 - \frac{1}{2}x + \frac{3}{8}x^2 - \frac{5}{16}x^3 + \cdots.$$

When $x = -\left(\dfrac{v^2}{c^2}\right)$, this becomes

$$\frac{1}{\sqrt{1-\frac{v^2}{c^2}}} = 1 + \frac{1}{2}\frac{v^2}{c^2} + \frac{3}{8}\frac{v^4}{c^4} + \frac{5}{16}\frac{v^6}{c^6} + \cdots.$$

Similarly, when $\alpha = 1/2$,

$$(1+x)^{1/2} = 1 + \frac{1}{2}x - \frac{1}{8}x^2 + \frac{1}{16}x^3 + \cdots \qquad \text{and}$$

$$\sqrt{1 - \frac{v^2}{c^2}} = 1 - \frac{1}{2}\frac{v^2}{c^2} - \frac{1}{8}\frac{v^4}{c^4} - \frac{1}{16}\frac{v^6}{c^6} + \cdots.$$

In the limit $v \ll c$, then,

$$1 - \sqrt{1 - \frac{v^2}{c^2}} \approx \frac{1}{2}\frac{v^2}{c^2} + \frac{1}{8}\frac{v^4}{c^4}, \qquad \frac{1}{\sqrt{1 - \frac{v^2}{c^2}}} - 1 \approx \frac{1}{2}\frac{v^2}{c^2} + \frac{3}{8}\frac{v^4}{c^4}.$$

Note that the v^6/c^6 and higher-order terms have been neglected; in practice, the v^4/c^4 terms are seldom used.

II - Algebraic

Consider the difference

$$1 - \sqrt{1 - \frac{v^2}{c^2}} = \left(1 - \sqrt{1 - \frac{v^2}{c^2}}\right) \times \frac{1 + \sqrt{1 - \frac{v^2}{c^2}}}{1 + \sqrt{1 - \frac{v^2}{c^2}}}$$

$$= \frac{1 - \left(\sqrt{1 - \frac{v^2}{c^2}}\right)^2}{1 + \sqrt{1 - \frac{v^2}{c^2}}}$$

$$= \frac{\frac{v^2}{c^2}}{1 + \sqrt{1 - \frac{v^2}{c^2}}}.$$

The denominator is seen to approach 2 as $v \ll c$, and so

$$1 - \sqrt{1 - \frac{v^2}{c^2}} \approx \frac{1}{2}\frac{v^2}{c^2}$$

for low speed. Similarly,

$$\frac{1}{\sqrt{1 - \frac{v^2}{c^2}}} - 1 = \frac{1 - \sqrt{1 - \frac{v^2}{c^2}}}{\sqrt{1 - \frac{v^2}{c^2}}} \approx \frac{1}{2}\frac{v^2}{c^2}$$

for low speed, where the previous result has been used.

Finding higher-order corrections by this method is possible, only slightly tedious, but fairly unenlightening. For example, for the next order, consider

$$\left(1 - \frac{1}{2}\frac{v^2}{c^2} - \sqrt{1 - \frac{v^2}{c^2}}\right) \times \frac{1 - \frac{1}{2}\frac{v^2}{c^2} + \sqrt{1 - \frac{v^2}{c^2}}}{1 - \frac{1}{2}\frac{v^2}{c^2} + \sqrt{1 - \frac{v^2}{c^2}}}.$$

This algebraic method is equivalent to that used to find a derivative of a square root by taking a limit.

III - Taylor Series

Letting $f(x) = (1+x)^{-1/2}$, $f(0) = 1$ and direct calculations give $f'(0) = -1/2$ and $f''(0) = 3/4$ (a generalization is not hard to do explicitly). Thus,

$$f(x) = \frac{1}{\sqrt{1+x}} \approx 1 + \left(-\frac{1}{2}\right)x + \frac{1}{2}\left(\frac{3}{4}\right)x^2 = 1 - \frac{1}{2}x + \frac{3}{8}x^2.$$

This is seen to be identical (when higher-order terms are computed) to that found by the binomial theorem, and letting $x = -v^2/c^2$ reproduces the previous result. Similary,

$$\sqrt{1+x} = (1+x)^{1/2} \approx 1 + \frac{1}{2}x - \frac{1}{8}x^4,$$

as before.

IV - Use the Machine

The mechanics of finding Taylor Series might often be left to mechanical devices. The following Maple commands reproduce the above results easily and almost immediately.

```
>g:=sqrt(1-(v/c)^2);
>series(g,v=0,8);
>series(1/g,v=0,8);
```

In the "`series`" commands above, the last argument is the order to which the series are calculated, and may be changed as desired (default is 6). Since the functions considered are even in v/c, the order is not the same as the number of terms.

1-1: All else being the same, including the rates of the chemical reactions that govern our brains and bodies, relativisitic phenomena would be more conspicuous if the speed of light were smaller. If we could attain the absolute speeds obtainable to us in the universe as it is, but with the speed of light being smaller, we would be able to move at speeds that would correspond to larger fractions of the speed of light, and in such instances relativistic effects would be more conspicuous.

1-3: Even if the judges would allow it, the observers in the moving spaceship would measure a longer time, since they would see the runners being timed by clocks that appear to run slowly compared to the ship's clocks. Actually, when the effects of length contraction are included (discussed in Section 1.4 and Appendix I), the runner's *speed* may be greater than, less than, or the same as that measured by an observer on the ground.

1-5: Note that the nonrelativistic approximation is not valid, as $v/c = 2/3$.

(a) See Example 1.1. In Equation (1.3), with t representing both the time measured by A *and* the time as measured in A's frame for the clock in B's frame to advance by t_0, we need

$$t - t_0 = t\left(1 - \sqrt{1 - \frac{v^2}{c^2}}\right) = t\left(1 - \sqrt{1 - \left(\frac{2}{3}\right)^2}\right) = t\,(0.255) = 1.00 \text{ s},$$

from which $t = 3.93$ s.

(b) A moving clock always seems to run slower. In this problem, the time t is the time that observer A measures as the time that B's clock takes to record a time change of t_0.

1-7: From Equation (1.3), for the time t on the earth to correspond to twice the time t_0 elapsed on the ship's clock,

$$\sqrt{1 - \frac{v^2}{c^2}} = \frac{1}{2}, \quad \text{so}$$

$$v = \frac{\sqrt{3}}{2}c = \frac{\sqrt{3}}{2}(2.998 \times 10^8 \text{ m/s}) = 2.60 \times 10^8 \text{ m/s},$$

retaining three significant figures.

1-9: The lifetime of the particle is t_0, and the distance the particle will travel is, from Equation (1.3),

$$vt = \frac{v t_0}{\sqrt{1 - \frac{v^2}{c^2}}} = \frac{(0.99)(2.998 \times 10^8 \text{ m/s})(1.00 \times 10^{-7} \text{ s})}{\sqrt{1 - (0.99)^2}} = 210 \text{ m}$$

to two significant figures.

1-11: See Example 1.3; for the intermediate calculations, note that

$$\lambda = \frac{c}{\nu} = \frac{c}{\nu_0}\frac{\nu_0}{\nu} = \lambda_0 \sqrt{\frac{1-v/c}{1+v/c}},$$

where the sign convention for v is that of Equation (1.8), with v positve for an approaching source and v negative for a receding source.

For this problem,

$$\frac{v}{c} = -\frac{(1.50 \times 10^4 \text{ km/s})(10^3 \text{ m/km})}{(2.998 \times 10^8 \text{ m/s})} = -0.0500,$$

so that

$$\lambda = \lambda_0 \sqrt{\frac{1-v/c}{1+v/c}} = (550 \text{ nm})\sqrt{\frac{1+0.0500}{1-0.0500}} = 578 \text{ nm}.$$

1-13: This problem may be done in several ways, all of which need to use the fact that when the frequencies due to the classical and relativistic effects are found, those frequencies, while differing by 1 Hz, will both be sufficiently close to $\nu_0 = 10^9$ Hz so that ν_0 could be used for an approximation to either.

In Equation (1.4), we have $v = 0$ and $V = -u$, where u is the speed of the spacecraft, moving away from the earth ($V < 0$). In Equation (1.6), we have $v = u$ (or $v = -u$ in Equation (1.8)). The classical and relativistic frequencies, ν_c and ν_r respectively, are

$$\nu_c = \frac{\nu_0}{1+(u/c)}, \qquad \nu_r = \nu_0 \sqrt{\frac{1-(u/c)}{1+(u/c)}} = \nu_0 \frac{\sqrt{1-(u^2/c^2)}}{1+(u/c)}.$$

The last expression for ν_r is motivated by the derivation of Equation (1.6), which essentially incorporates the classical result (counting the number of ticks), and allows expression of the ratio

$$\frac{\nu_c}{\nu_r} = \frac{1}{\sqrt{1-(u^2/c^2)}}.$$

Use of the above forms for the frequencies allows the calculation of the ratio

$$\frac{\Delta \nu}{\nu_0} = \frac{\nu_c - \nu_r}{\nu_0} = \frac{1-\sqrt{1-(u^2/c^2)}}{1+(u/c)} = \frac{1 \text{ Hz}}{10^9 \text{ Hz}} = 10^{-9}.$$

Attempts to solve this equation exactly are not likely to be met with success, and even numerical solutions would require a higher precision than is commonly available. However, recognizing that the numerator $1 - \sqrt{1-(u^2/c^2)}$ is of the form that can be approximated using the methods outlined at the beginning of this chapter, we can use $1 - \sqrt{1-(u^2/c^2)} \approx (1/2)(u^2/c^2)$. The denominator will be indistinguishable from 1 at low speed, with the result

$$\frac{1}{2}\frac{u^2}{c^2} = 10^{-9},$$

which is solved for

$$u = \sqrt{2 \times 10^{-9}}\, c = 1.34 \times 10^4 \text{ m/s} = 13.4 \text{ km/s}.$$

Similar to what was done at the beginning of this chapter, the Taylor series for the desired function of u can be found by a computer. The Maple commands would be

```
>f:=(1-sqrt(1-u^2))/(1+u);
>series(f,u=0);
```

(note that for these commands, "u" represents the ratio of the recessional speed to the speed of light).

Mention had been made above of the limited possibility of a numerical solution. Depending on which release of Maple is used, a numerical solution is indeed possible. Maple 7 will solve the given equation with the command

```
>solve(f=1E-9);
```

with the results .00004472235955, −.00004472035955 for u/c (Maple will give both positive and negative roots, and we need to recognize which we want, as well as the limitation on precision).

1-15: The transverse Doppler effect corresponds to a direction of motion of the light source that is perpendicular to the direction from it to the observer; the angle $\theta = \pm\frac{\pi}{2}$ (or $\pm 90°$), so $\cos\theta = 0$, and $\nu = \nu_0\sqrt{1-v^2/c^2}$, which is Equation (1.5).

For a receding source, $\theta = \pi$ (or $180°$), and $\cos\theta = -1$. The given expression becomes

$$\nu = \nu_0 \frac{\sqrt{1-\frac{v^2}{c^2}}}{1+\frac{v}{c}} = \nu_0 \frac{\sqrt{1-\frac{v}{c}}\sqrt{1+\frac{v}{c}}}{\sqrt{1+\frac{v}{c}}\sqrt{1+\frac{v}{c}}} = \nu_0\sqrt{\frac{1-v/c}{1+v/c}},$$

which is Equation (1.6).

Relativity

For an approaching source, $\theta = 0$, $\cos\theta = 1$, and the given expression becomes

$$\nu = \nu_0 \frac{\sqrt{1-\frac{v^2}{c^2}}}{1-\frac{v}{c}} = \nu_0 \frac{\sqrt{1+\frac{v}{c}}\sqrt{1-\frac{v}{c}}}{\sqrt{1-\frac{v}{c}}\sqrt{1-\frac{v}{c}}} = \nu_0 \sqrt{\frac{1+v/c}{1-v/c}},$$

which is Equation (1.7).

1-17: The astronaut's proper length (height) is 6 ft, and this is what any observer in the spacecraft will measure. From Equation (1.9), an observer on the earth would measure

$$L = L_0 \sqrt{1-\frac{v^2}{c^2}} = (6 \text{ ft}) \sqrt{1-(0.90)^2} = 2.6 \text{ ft}.$$

1-19: The time will be the length as measured by the observer divided by the speed, or

$$t = \frac{L}{v} = \frac{L_0 \sqrt{1-\frac{v^2}{c^2}}}{v} = \frac{(1.00 \text{ m})\sqrt{1-(0.100)^2}}{(0.100)(2.998 \times 10^8 \text{ m/s})} = 3.32 \times 10^{-8} \text{ s}.$$

1-21: If the antenna has a length L' as measured by an observer on the spacecraft (L' is *not* either L or L_0 in Equation (1.9)), the projection of the antenna onto the spacecraft will have a length $L'\cos(10°)$, and the projection onto an axis perpendicular to the spacecraft's axis will have a length $L'\sin(10°)$. To an observer on the earth, the length in the direction of the spacecraft's axis will be contracted as described by Equation (1.9), while the length perpendicular to the spacecraft's motion will appear unchanged. The angle as seen from the earth will then be

$$\arctan\left[\frac{L'\sin(10°)}{L'\cos(10°)\sqrt{1-\frac{v^2}{c^2}}}\right] = \arctan\left[\frac{\tan(10°)}{\sqrt{1-(0.70)^2}}\right] = 14°.$$

The generalization of the above is that if the angle is θ_0 as measured by an observer on the spacecraft, an observer on the earth would measure an angle θ given by

$$\tan\theta = \frac{\tan\theta_0}{\sqrt{1-\frac{v^2}{c^2}}}.$$

1-23: The age difference will be the difference in the times that each measures the round trip to take, or

$$\Delta t = 2\frac{L_0}{v}\left(1 - \sqrt{1 - \frac{v^2}{c^2}}\right) = 2\frac{4\text{ yr}}{0.9}\left(1 - \sqrt{1 - (0.9)^2}\right) = 5\text{ yr}.$$

1-25: It is convenient to maintain the relationship from Newtonian mechanics, in that a force on an object changes the object's momentum; symbolically, $\mathbf{F} = \dfrac{d\mathbf{p}}{dt}$ should still be valid. In the absence of forces, momentum should be conserved in any inertial frame, and the conserved quantity is $\mathbf{p} = \gamma m\mathbf{v}$, not $m\mathbf{v}$.

1-27: For a given mass M, the ratio of the mass liberated to the mass energy is

$$\frac{M \times (5.4 \times 10^6\text{ J/kg})}{M \times (2.998 \times 10^8\text{ m/s})^2} = 6.0 \times 10^{-11}.$$

1-29: If the kinetic energy $\text{KE} = E_0 = mc^2$, then $E = 2\,mc^2$ and Equation (23) reduces to

$$\frac{1}{\sqrt{1 - \dfrac{v^2}{c^2}}} = 2$$

($\gamma = 2$ in the notation of Section 1.7). Solving for v,

$$v = \frac{\sqrt{3}}{2}c = 2.60 \times 10^8\text{ m/s}.$$

1-31: Classically,

$$v = \sqrt{\frac{2\,\text{KE}}{m_e}} = \sqrt{\frac{2(0.100\text{ MeV})(1.602 \times 10^{-19}\text{ J/eV})}{(9.1095 \times 10^{-31}\text{ kg})}} = 1.88 \times 10^8\text{ m/s}.$$

Relativistically, solving Equation (1.23) for v as a function of KE,

$$v = c\sqrt{1 - \left(\frac{m_e c^2}{E}\right)^2} = c\sqrt{1 - \left(\frac{m_e c^2}{m_e c^2 + \text{KE}}\right)^2}$$

$$= c\sqrt{1 - \left(\frac{1}{1 + (\text{KE}/(m_e c^2))}\right)^2}.$$

With $\text{KE}/(m_e c^2) = (0.100\text{ MeV})/(0.511\text{ MeV}) = 0.100/0.511$,

$$v = (2.998 \times 10^8\text{ m/s})\sqrt{1 - \left(\frac{1}{1 + (0.100/0.511)}\right)^2} = 1.64 \times 10^8\text{ m/s}.$$

The two speeds are comparable, but not the same; for larger values of the ratio of the kinetic and rest energies, larger discrepancies would be found.

1-33: Using Equation (1.22) in Equation (1.23) and solving for $\dfrac{v}{c}$,

$$\frac{v}{c} = \sqrt{1 - \left(\frac{E_0}{E}\right)^2} = \left[1 - \left(\frac{E_0}{E}\right)^2\right]^{1/2}.$$

With $E = 21\,E_0$, that is, $E = E_0 + (20)\,E_0$,

$$v = c\sqrt{1 - \left(\frac{1}{21}\right)^2} = 0.9989\,c.$$

(This is consistent with the expression derived in Problem 1-32.)

1-35: The difference in energies will be, from Equation (1.23),

$$m_e c^2 \left[\frac{1}{\sqrt{1 - (v_2/c)^2}} - \frac{1}{\sqrt{1 - (v_1/c)^2}}\right]$$

$$= (0.511 \text{ MeV}) \left[\frac{1}{\sqrt{1 - (2.4/3.0)^2}} - \frac{1}{\sqrt{1 - (1.2/3.0)^2}}\right] = 0.294 \text{ MeV},$$

to three significant figures.

1-37: Using the expression in Equation (1).20 for the kinetic energy, the ratio of the two quantities is

$$\frac{\tfrac{1}{2}\gamma m v^2}{\text{KE}} = \frac{1}{2}\frac{v^2}{c^2}\left(\frac{\gamma}{\gamma - 1}\right) = \frac{1}{2}\frac{v^2}{c^2}\left[\frac{1}{1 - \sqrt{1 - \dfrac{v^2}{c^2}}}\right].$$

Algebraically, this quantity is not equal to 1 except at $v = 0$. For low speeds, $v \ll c$, the quantity in square brackets is approximately $\dfrac{1}{2}\dfrac{v^2}{c^2}$ (see the text at the end of Section 1.8 or the beginning of this chapter), reflecting the fact that the classical and relativistic kinetic energies have the same form in the nonrelativistic limit. However, as $v \to c$ (or $\gamma \to \infty$), the expressions are not the same, even though both $\tfrac{1}{2}\gamma m v^2$ and $\text{KE} = (\gamma - 1)\,m c^2$ become infinitely large. To see this explicitly, note that the ratio $\left(\dfrac{\gamma}{\gamma - 1}\right) \to 1$ as $\gamma \to \infty$, so that the expression approaches $\tfrac{1}{2}$ as $v \to c$. This is consistent with setting $v = c$ in the last expression on the right above.

1-39: Measured from the original center of the box, so that the original position of the center of mass is 0, the final position of the center of mass is

$$\left(\frac{M}{2} - m\right)\left(\frac{L}{2} + S\right) - \left(\frac{M}{2} + m\right)\left(\frac{L}{2} - S\right) = 0.$$

Expanding the products and cancelling similar terms ($\frac{M}{2}\frac{L}{2}$, mS), the result $MS = mL$ is obtained. The distance S is the product vt, where, as shown in the problem statement, $v \approx E/Mc$ (approximate in the nonrelativistic limit $M \gg E/c^2$) and $t \approx L/c$. Then,

$$m = \frac{MS}{L} = \frac{M}{L}\frac{E}{Mc}\frac{L}{c} = \frac{E}{c^2}.$$

1-41: To cross the galaxy in a matter of minutes, the proton must be highly relativistic, with $v \approx c$ (but $v < c$, of course). The energy of the proton will be $E = E_0\gamma$, where E_0 is the proton's rest energy and $\gamma = 1/\sqrt{1 - \frac{v^2}{c^2}}$. However, γ, from Equation (1.9), is the same as the ratio L_0/L, where L is the diameter of the galaxy in the proton's frame of reference, and for the highly-relativistic proton $L \approx ct$, where t is the time in the proton's frame that it takes to cross the galaxy. Combining,

$$E = E_0\gamma = E_0\frac{L_0}{L} \approx E_0\frac{L_0}{ct} \approx (10^9 \text{ eV})\frac{(10^5 \text{ ly})}{c(300 \text{ s})} \times (3 \times 10^7 \text{ s/yr}) = 10^{19} \text{ eV}.$$

1-43: Taking magnitudes in Equation (1.16),

$$p = \frac{m_e v}{\sqrt{1 - \frac{v^2}{c^2}}} = \frac{(0.511 \text{ MeV}/c^2)(0.600\,c)}{\sqrt{1 - (0.600)^2}} = 0.383 \text{ MeV}/c.$$

1-45: When the kinetic energy of an electron is equal to its rest energy, the total energy is twice the rest energy, and Equation (1.24) becomes

$$4m_e^2 c^4 = m_e^2 c^4 + p^2 c^2, \quad \text{or} \quad p = \sqrt{3}\,(m_e c^2)/c = \sqrt{3}\,(511 \text{ keV}/c) = 885 \text{ keV}/c.$$

The result of Problem 1-29 could be used directly; $\gamma = 2$, $v = (\sqrt{3}/2)c$, and Equation (1.17) gives $p = \sqrt{3}\,m_e c$, as above.

1-47: Solving Equation (1.23) for the speed v in terms of the rest energy E_0 and the total energy E,

$$v = c\sqrt{1 - \left(\frac{E_0}{E}\right)^2} = c\sqrt{1 - \left(\frac{0.938}{3.500}\right)^2} = 0.963\,c,$$

numerically 2.888×10^8 m/s. (The result of Problem 1-32 does *not* give an answer accurate to three significant figures.) The value of the speed may be substituted into Equation (1.16) (or the result of Problem 1-46), or Equation (1.24) may be solved for the magnitude of the momentum,

$$p = \sqrt{(E/c)^2 - (E_0/c)^2} = \sqrt{(3.500\text{ GeV}/c)^2 - (0.93828\text{ GeV}/c)^2} = 3.372\text{ GeV/c}.$$

(Although the final result is not affected, a more precise value for the proton rest mass, taken from the front endpapers, was used in the last calculation.)

1-49: From $E = mc^2 + \text{KE}$ and Equation (1.24),

$$\left(mc^2 + \text{KE}\right)^2 = m^2c^4 + p^2c^2.$$

Expanding the binomial, cancelling the m^2c^4 term, and solving for m,

$$m = \frac{(pc)^2 - \text{KE}^2}{2c^2\,\text{KE}} = \frac{(335\text{ MeV})^2 - (62\text{ MeV})^2}{2c^2\,(62\text{ MeV})} = 874\text{ MeV}/c^2.$$

The particle's speed may be found any number of ways; a very convenient result is that of Problem 1-46, giving

$$v = c^2\frac{p}{E} = c\frac{pc}{mc^2 + \text{KE}} = c\frac{335\text{ MeV}}{874\text{ MeV} + 62\text{ MeV}} = 0.36\,c.$$

There's a neat algebraic "trick" that may be used in this and many similar problems. (In what follows, factors of c will not be included.) Essentially, the problem reduces mathematically to solving the two equations

$$E = m + \text{KE}, \qquad E^2 = m^2 + p^2$$

for E and m, given known values for p and KE. Rewrite the two equations as

$$E - m = \text{KE}, \qquad E^2 - m^2 = (E-m)(E+m) = p^2$$

and substitute the first into the second to obtain $E + m = \dfrac{p^2}{\text{KE}}$ (the KE = 0 case is trivial). Adding this to $E - m = \text{KE}$, and then subtracting the same relation gives

$$E = \frac{p^2 + \text{KE}^2}{2\,\text{KE}}, \qquad m = \frac{p^2 - \text{KE}^2}{2\,\text{KE}},$$

as obtained above.

1-51: The given observation that the two explosions occur at the same place to the second observer means that $x' = 0$ in Equation (1.41), and so the second observer is moving at a speed

$$v = \frac{x}{t} = \frac{1.00 \times 10^5 \text{ m}}{2.00 \times 10^{-3} \text{ s}} = 5.00 \times 10^7 \text{ m/s}$$

with respect to the first observer. Inserting this into Equation (1.44),

$$t' = \frac{t - \frac{x^2}{tc^2}}{\sqrt{1 - (x/ct)^2}} = t\frac{1 - \frac{x^2}{c^2 t^2}}{\sqrt{1 - \frac{x^2}{c^2 t^2}}} = t\sqrt{1 - \frac{(x/t)^2}{c^2}}$$

$$= (2.00 \text{ ms})\sqrt{1 - \frac{(5.00 \times 10^7 \text{ m/s})^2}{(2.998 \times 10^8 \text{ m/s})^2}} = 1.97 \text{ ms}.$$

(For this calculation, the approximation $\sqrt{1 - (x/ct)^2} \approx 1 - (x^2/2c^2 t^2)$ is valid to three significant figures.)

An equally valid method, and a good check, is to note that when the relative speed of the observers (5.00×10^7 m/s) has been determined, the time interval that the second observer measures should be that given by Equation (1.3) (but be careful of which time it t, which is t_0). Algebraically and numerically, the different methods give the same result.

1-53: (a) A convenient choice for the origins of both the unprimed and primed coordinate systems is the point, in both space and time, where the ship receives the signal. Then, in the unprimed frame (given here as the frame of the fixed stars, one of which may be the source), the signal was sent at a time $t = -r/c$, where r is the distance from the source to the place where the ship receives the signal, and the minus sign merely indicates that the signal was sent before it was received.

Take the direction of the ship's motion (assumed parallel to its axis) to be the positive x-direction, so that in the frame of the fixed stars (the unprimed frame), the signal arrives at an angle θ with respect to the positive x-direction. In the unprimed frame, $x = r\cos\theta$ and $y = r\sin\theta$. From Equation (1.41),

$$x' = \frac{x - vt}{\sqrt{1 - \frac{v^2}{c^2}}} = \frac{r\cos\theta - (-r/c)}{\sqrt{1 - \frac{v^2}{c^2}}} = r\frac{\cos\theta + (v/c)}{\sqrt{1 - \frac{v^2}{c^2}}},$$

and $y' = y = r\sin\theta$. Then,

$$\tan\theta' = \frac{y'}{x'} = \frac{\sin\theta}{(\cos\theta + (v/c))\Big/\sqrt{1-\frac{v^2}{c^2}}}, \quad \text{and}$$

$$\theta' = \arctan\left[\frac{\sin\theta\sqrt{1-\frac{v^2}{c^2}}}{\cos\theta + (v/c)}\right].$$

(b) From the form of the result of part (a), it can be seen that the numerator of the term in square brackets is less than $\sin\theta$, and the denominator is greater than $\cos\theta$, and so $\tan\theta' < \tan\theta$ and $\theta' < \theta$ when $v \neq 0$. Looking out of a porthole, the sources, including the stars, will appear to be in directions closer to the direction of the ship's motion than they would for a ship with $v = 0$. As $v \to c$, $\theta' \to 0$, and all stars appear to be almost on the ship's axis (farther forward in the field of view).

1-55: (a) If the man on the moon sees A approaching with speed $v = 0.800\,c$, then the observer on A will see the man in the moon approaching with speed $v = 0.800\,c$. The relative velocities will have opposite directions, but the relative speeds will be the same. The speed with which B is seen to approach A, to an observer in A, is then

$$\frac{0.800 + 0.900}{1 + (0.800)(0.900)}c = 0.988\,c.$$

(b) Similarly, the observer on B will see the man on the moon approaching with speed $0.900\,c$, and the apparent speed of A, to an observer on B, will be

$$\frac{0.900 + 0.800}{1 + (0.900)(0.800)}c = 0.988\,c.$$

(Note that Equation (1.49) is unchanged if V'_x and v are interchanged.)

Chapter 2 - Particle Properties of Waves

2-1: Planck's constant gives a measure of the energy at which quantum effects are observed. If Planck's constant had a smaller value, while all other physical quantities, such as the speed of light, remained the same, quantum effects would be seen for phenomena that occur at higher frequncies or shorter wavelengths. That is, quantum phenomena would be less conspicuous than they are now.

2-3: No; the relation is given in Equation (2.8) and Equation (2.9),

$$\text{KE}_{max} = h\nu - \phi = h\left(\nu - \nu_0\right),$$

so that while KE_{max} is a *linear* function of the frequency ν of the incident light, KE_{max} is not *proportional* to the frequency.

2-5: From Equation (2.11),

$$E = \frac{1.240 \times 10^{-6} \text{ eV·m}}{700 \times 10^{-9} \text{ m}} = 1.77 \text{ eV}.$$

Or, in terms of joules,

$$E = \frac{\left(6.626 \times 10^{-34} \text{ J·s}\right)\left(2.998 \times 10^8 \text{ m/s}\right)}{700 \times 10^{-9} \text{ m}} = 2.84 \times 10^{-19} \text{ J}.$$

2-7: The number of photons per unit time is the total energy per unit time (the power) divided by the energy per photon, or

$$\frac{P}{E} = \frac{P}{h\nu} = \frac{1.00 \times 10^3 \text{ J/s}}{\left(6.626 \times 10^{-34} \text{ J·s}\right)\left(880 \times 10^3 \text{ Hz}\right)} = 1.72 \times 10^{30} \text{ photons/s}.$$

2-9: (a) The number of photons per unit time per unit area will be the energy per unit time per unit area (the power per unit area, P/A), divided by the energy per photon, or

$$\frac{(P/A)}{h\nu} = \frac{1.4 \times 10^3 \text{ W/m}^2}{\left(6.626 \times 10^{-34} \text{ J·s}\right)\left(5.0 \times 10^{14} \text{ Hz}\right)} = 4.2 \times 10^{21} \text{ photons/(s·m}^2).$$

(b) With the reasonable assumption that the sun radiates uniformly in all directions, all points at the same distance from the sun should have the same flux of energy, even if there is no surface to absorb the energy. The total power is then

$$(P/A)\, 4\pi R_{E-S}^2 = \left(1.4 \times 10^3 \text{ W/m}^2\right) 4\pi \left(1.5 \times 10^{11} \text{ m}\right)^2 = 4.0 \times 10^{26} \text{ W},$$

where R_{E-S} is the mean Earth-Sun distance, commonly abbreviated as "1 AU," for "astronomical unit." The number of photons emitted per second is this power divided by the energy per photon, or

$$\frac{4.0 \times 10^{26} \text{ J/s}}{(6.626 \times 10^{-34} \text{ J}\cdot\text{s})(5.0 \times 10^{14} \text{ Hz})} = 1.2 \times 10^{45} \text{ photons/s}.$$

(c) The photons are all moving at the same speed c, and in the same direction (spreading is not significant on the scale of the earth), and so the number of photons per unit time per unit area is the product of the number per unit volume and the speed. Using the result from part (a),

$$\frac{4.2 \times 10^{21} \text{ photons}/(\text{s}\cdot\text{m}^2)}{2.998 \times 10^8 \text{ m/s}} = 1.4 \times 10^{13} \text{ photons/m}^3.$$

2-11: Expressing Equation (2.9) in terms of $\lambda = c/\nu$ and $\lambda_0 = c/\nu_0$, and performing the needed algebraic manipulations,

$$\lambda = \frac{hc}{(hc/\lambda_0) + \text{KE}_{\max}} = \lambda_0 \left[1 + \frac{\text{KE}_{\max} \lambda_0}{hc}\right]^{-1}$$
$$= 230 \text{ nm} \left[1 + \frac{(1.5 \text{ eV})(230 \times 10^{-9} \text{ m})}{(1.240 \times 10^{-6} \text{ eV}\cdot\text{m})}\right]^{-1} = 180 \text{ nm}.$$

Note that in the above calcuation, λ_0 was used twice, once expressed in terms of nanometers and once in terms of meters, as convenient. Of course, whether or not the extra algebra used to save one calculational step is an advantage is subjective.

2-13: The maximum wavelength would correspond to the least energy that would allow an electron to be emitted, so the incident energy wold be equal to the work function, and

$$\lambda_{\max} = \frac{hc}{\phi} = \frac{1.240 \times 10^{-6} \text{ eV}\cdot\text{m}}{2.3 \text{ eV}} = 539 \text{ nm},$$

where the value of ϕ for sodium is taken from Table 2.1.

From Equation (2.8),

$$\text{KE}_{\max} = h\nu - \phi = \frac{hc}{\lambda} - \phi = \frac{1.240 \times 10^{-6} \text{ eV}\cdot\text{m}}{200 \times 10^{-9} \text{ m}} - 2.3 \text{ eV} = 3.9 \text{ eV}.$$

2-15: Because only 0.10% of the light creates photoelectrons, the available power is $(1.0 \times 10^{-3})(1.5 \times 10^{-3} \text{ W}) = 1.5 \times 10^{-6}$ W. The current will be the product of the number of photoelectrons per unit time and the electron charge, or

$$I = e\frac{P}{E} = e\frac{P}{hc/\lambda} = e\frac{P\lambda}{hc} = (1\text{ e})\frac{(1.5 \times 10^{-6} \text{ J/s})(400 \times 10^{-9} \text{ m})}{(1.240 \times 10^{-6} \text{ eV}\cdot\text{m})} = 0.48 \ \mu\text{A}.$$

In this calculation, note that the units of the result are J/(V·s), and that because one volt is one joule per coulomb, the answer has units of coulombs per second, or amperes.

2-17: Denoting the two energies and frequencies with subscripts 1 and 2,

$$\text{KE}_{\text{max},1} = h\nu_1 - \phi, \qquad \text{KE}_{\text{max},2} = h\nu_2 - \phi.$$

Subtracting to eliminate the work function ϕ and dividing by $\nu_1 - \nu_2$,

$$h = \frac{\text{KE}_{\text{max},2} - \text{KE}_{\text{max},1}}{\nu_2 - \nu_1} = \frac{19.7 \text{ eV} - 0.52 \text{ eV}}{12.0 \times 10^{14} \text{ Hz} - 8.5 \times 10^{14} \text{ Hz}} = 4.1 \times 10^{-15} \text{ eV}\cdot\text{s}$$

to the allowed two significant figures. Keeping an extra figure gives

$$h = 4.14 \times 10^{-15} \text{ eV}\cdot\text{s} = 6.64 \times 10^{-15} \text{ J}\cdot\text{s}.$$

The work function ϕ may be obtained by substituting the above result into either of the above expressions relating the frequencies and the energies, yielding $\phi = 3.0$ eV to the same two significant figures, or the equations may be solved by rewriting them as

$$\text{KE}_{\text{max},1}\nu_2 = h\nu_1\nu_2 - \phi\nu_2, \qquad \text{KE}_{\text{max},2}\nu_1 = h\nu_2\nu_1 - \phi\nu_1,$$

subtracting to eliminate the product $h\nu_1\nu_2$ and dividing by $\nu_1 - \nu_2$ to obtain

$$\phi = \frac{\text{KE}_{\text{max},2}\nu_1 - \text{KE}_{\text{max},1}\nu_2}{\nu_2 - \nu_1}$$
$$= \frac{(19.7 \text{ eV})(8.5 \times 10^{14} \text{ Hz}) - (0.52 \text{ eV})(12.0 \times 10^{14} \text{ Hz})}{(12.0 \times 10^{14} \text{ Hz} - 8.5 \times 10^{14} \text{ Hz})} = 3.0 \text{ eV}.$$

(This last calculation, while possibly more cumbersome than direct substitution, reflects the result of solving the system of equations using a symbolic-manipulation program; using such a program for this problem is, of course, a case of "swatting a fly with a sledgehammer".)

Particle Properties of Waves 17

2-19: Consider the proposed interaction in the frame of the electron initially at rest. The photon's initial momentum is $p_0 = E_0/c$, and if the electron were to attain all of the photon's momentum and energy, the final momentum of the electron must be $p_e = p_0 = p$, the final electron kinetic energy must be $KE = E_0 = pc$, and so the final electron energy is $E_e = pc + m_e c^2$. However, for any electron we must have $E_e^2 = (pc)^2 + (m_e c^2)^2$. Equating the two expressions for E_e^2,

$$E_e^2 = (pc)^2 + (m_e c^2)^2 = (pc + m_e c^2)^2 = (pc)^2 + 2(pc)(m_e c^2) + (m_e c^2)^2,$$

or

$$0 = 2(pc)(m_e c^2).$$

This is only possible if $p = 0$, in which case the photon had no initial momentum and no initial energy, and hence could not have existed.

To see the same result without using as much algebra, the electron's final kinetic energy is

$$\sqrt{p^2 c^2 + m_e^2 c^4} - m_e c^2 \neq pc$$

for nonzero p.

An easier alternative is to consider the interaction in the frame where the electron is at rest *after* absorbing the photon. In this frame, the final energy is the rest energy of the electron, $m_e c^2$, but before the interaction, the electron would have been moving (to conserve momentum), and hence would have had more energy than after the interaction, and the photon would have had positive energy, so energy could not be conserved.

2-21: For the highest frequency, the electrons will acquire all of their kinetic energy from the accelerating voltage, and this energy will appear as the electromagnetic radiation emitted when these electrons strike the screen. The frequency of this radiation will be

$$\nu = \frac{E}{h} = \frac{eV}{h} = \frac{(1\text{ e})(10 \times 10^3 \text{ V})}{(4.136 \times 10^{-15} \text{ eV·s})} = 2.4 \times 10^{18} \text{ V},$$

which corresponds to x-rays.

2-23: Solving Equation (2.13) for θ with $n = 1$,

$$\theta = \arcsin\left(\frac{\lambda}{2d}\right) = \arcsin\left(\frac{0.030 \text{ nm}}{2(0.300 \text{ nm})}\right) = 2.9°.$$

Extra Expressions for Compton Effect Problems

Many of the problems from this section of the text involve the recoil direction of the electron in the Compton effect. A commonly-occurring relation is obtained from the two displayed equations preceding Equation (2.18);

$$pc \cos\theta = h\nu - h\nu' \cos\phi$$

$$pc \sin\theta = h\nu' \sin\phi.$$

Dividing the second by the first yields

$$\tan\theta = \frac{\nu' \sin\phi}{\nu - \nu' \cos\phi} = \frac{\sin\phi}{(\nu/\nu') - \cos\phi} = \frac{\sin\phi}{(\lambda'/\lambda) - \cos\phi}.$$

Rewriting $\lambda'/\lambda = (\lambda + \Delta\lambda)/\lambda = 1 + (\Delta\lambda/\lambda)$ gives

$$\tan\theta = \frac{\sin\phi}{(\Delta\lambda/\lambda) + (1 - \cos\phi)}.$$

Further algebraic and trigonometric manipulations are possible, and will be demonstrated as appropriate for the specific situations.

2-25: From Equation (2.15),

$$\nu = \frac{cp}{h} = \frac{(2.998 \times 10^8 \text{ m/s})(1.1 \times 10^{-23} \text{ kg·m/s})}{(6.626 \times 10^{-34} \text{ J·s})} = 5.0 \times 10^{18} \text{ Hz}.$$

2-27: Following the steps that led to Equation (2.22), but with a sodium atom instead of an electron,

$$\lambda_{C,\text{Na}} = \frac{h}{cM_{\text{NA}}} = \frac{(6.626 \times 10^{-34} \text{ J·s})}{(2.998 \times 10^8 \text{ m/s})(3.82 \times 10^{-26} \text{ kg})} = 5.8 \times 10^{-17} \text{ m},$$

or 5.8×10^{-8} nm, which is much less than 0.1 nm. (Here, the rest mass $M_{\text{NA}} = 3.82 \times 10^{-26}$ kg was taken from Problem 2-24.)

2-29: Solving Equation (2.23) for λ, the wavelength of the x-rays in the direct beam,

$$\lambda = \lambda' - \lambda_C(1 - \cos\phi) = 2.2 \text{ pm} - (2.426 \text{ pm})(1 - \cos 45°) = 1.5 \text{ pm}$$

to the given two significant figures.

Particle Properties of Waves

2-31: Rewriting Equation (2.23) in terms of frequencies, with $\lambda = c/\nu$ and $\lambda' = c/\nu'$, and with $\cos 90° = 0$,

$$\frac{c}{\nu'} = \frac{c}{\nu} + \lambda_C$$

and solving for ν' gives

$$\nu' = \left[\frac{1}{\nu} + \frac{\lambda_C}{c}\right]^{-1} = \left[\frac{1}{3.0 \times 10^{19} \text{ Hz}} + \frac{2.426 \times 10^{-12} \text{ m}}{2.998 \times 10^8 \text{ m/s}}\right]^{-1} = 2.4 \times 10^{19} \text{ Hz}.$$

The above method avoids the intermediate calculation of wavelengths.

2-33: Solving Equation (2.23) for $\cos \phi$,

$$\cos \phi = 1 + \frac{\lambda}{\lambda_C} - \frac{\lambda'}{\lambda_C} = 1 + \left(\frac{mc^2}{E} - \frac{mc^2}{E'}\right) = 1 + \left(\frac{511 \text{ keV}}{100 \text{ keV}} - \frac{511 \text{ keV}}{90 \text{ keV}}\right) = 0.432,$$

from which $\phi = 64°$ to two significant figures.

2-35: For the electron to have the maximum recoil energy, the scattering angle must be $180°$, and Equation (2.20) becomes $mc^2 \text{KE}_{\max} = 2(h\nu)(h\nu')$, where $\text{KE}_{\max} = (h\nu - h\nu')$ has been used. To simplify the algebra somewhat, conisder

$$\nu' = \nu \frac{\lambda}{\lambda'} = \frac{\nu}{1 + (\Delta\lambda/\lambda)} = \frac{\nu}{1 + (2\lambda_C/\lambda)} = \frac{\nu}{1 + 2\nu\lambda_C/c},$$

where $\Delta\lambda = 2\lambda_C$ for $\phi = 180°$. With this expression,

$$\text{KE}_{\max} = \frac{2(h\nu)(h\nu')}{mc^2} = \frac{2(h\nu)^2/(mc^2)}{1 + (2\nu\lambda_C/c)}.$$

Using $\lambda_C = h/(mc)$ (which is Equation (2.22)) gives the desired result.

2-37: As presented in the text, the energy of the scattered photon is known in terms of the scattered angle, not the recoil angle of the scattering electron. Consider the expression for the recoil angle as given preceding the solution to Problem 2-25:

$$\tan \theta = \frac{\sin \phi}{(\Delta\lambda/\lambda) + (1 - \cos \phi)} = \frac{\sin \phi}{(\lambda_C/\lambda)(1 - \cos \phi) + (1 - \cos \phi)}$$

$$= \frac{\sin \phi}{\left(1 + \frac{\lambda_C}{\lambda}\right)(1 - \cos \phi)}.$$

For the given problem, with $E = mc^2$, $\lambda = hc/E = h/(mc) = \lambda_C$, so the above expression reduces to
$$\tan\theta = \frac{\sin\phi}{2(1-\cos\phi)}.$$

At this point, there are many ways to proceed; a numerical solution with $\theta = 40°$ gives $\phi = 61.6°$ to three significant figures. For an analytic solution which avoids the intermediate calculation of the scattering angle ϕ, one method is to square both sides of the above relation and use the trigonometric identity $\sin^2\phi = 1 - \cos^2\phi = (1 + \cos\phi)(1 - \cos\phi)$ to obtain

$$4\tan^2\theta = \frac{1+\cos\phi}{1-\cos\phi}$$

(the factor $1 - \cos\phi$ may be divided, as $\cos\phi = 1$, $\phi = 0$, represents an undeflected photon, and hence no interaction). This may be re-expressed as

$$(1-\cos\phi)(4\tan^2\theta) = 1+\cos\phi = 2 - (1-\cos\phi), \quad \text{or}$$

$$1-\cos\phi = \frac{2}{1+4\tan^2\theta}, \quad 2-\cos\phi = \frac{3+4\tan^2\theta}{1+4\tan^2\theta}.$$

Then, with $\lambda' = \lambda + \lambda_C(1 - \cos\phi) = \lambda_C(2 - \cos\phi)$,

$$E' = E\frac{\lambda}{\lambda'} = E\frac{1+4\tan^2\theta}{3+4\tan^2\theta} = (511\text{ keV})\frac{1+4\tan^2(40°)}{3+4\tan^2(40°)} = 335\text{ eV}.$$

An equivalent but slightly more cumbersome method is to use the trigonometric identities
$$\sin\phi = 2\sin\frac{\phi}{2}\cos\frac{\phi}{2}, \quad 1-\cos\phi = 2\sin^2\frac{\phi}{2}$$
in the expression for $\tan\theta$ to obtain
$$\tan\theta = \frac{1}{2}\cot\frac{\phi}{2}, \quad \phi = 2\arctan\left(\frac{1}{2\tan\theta}\right)$$
yielding the result $\phi = 61.6°$ more readily.

The above expression for ϕ in terms of θ is that obtained from MAPLE, using the single command
```
>solve(tan(theta)=sin(phi)/2/(1-cos(phi)),phi);
```
It's worth noting that the above analytic method may be extended to any value of the ratio λ_C/λ, with a corresponding increase in the complexity of the expressions.

Particle Properties of Waves 21

2-39: The energy of each photon will be the sum of one particle's rest and kinetic energies, 1.511 MeV (keeping an extra significant figure). The wavelength of each photon will be

$$\lambda = \frac{hc}{E} = \frac{1.240 \times 10^{-6} \text{ eV}\cdot\text{m}}{1.511 \times 10^6 \text{ eV}} = 8.21 \times 10^{-13} \text{ m} = 0.821 \text{ pm}.$$

2-41: Following the hint,

$$\lambda_C = \frac{h}{mc} = \frac{2hc}{2mc^2} = \frac{2hc}{E_{\min}},$$

where $E_{\min} = 2mc^2$ is the minimum photon energy needed for pair production. The scattered wavelength (a *maximum*) corresponding to this minimum energy is $\lambda'_{\max} = (h/E_{\min})$, so $\lambda_C = 2\lambda'_{\max}$.

At this point, it is possible to say that for the most energetic incoming photons, $\lambda \sim 0$, and so $1 - \cos\phi = \frac{1}{2}$ for $\lambda' = \lambda_C/2$, from which $\cos\phi = \frac{1}{2}$ and $\phi = 60°$. As an alternative, the angle at which the scattered photons will have wavelength $\lambda'_{\max}$ can be found as a function of the incoming photon energy E; solving Equation (2.23) with $\lambda' = \lambda'_{\max}$,

$$\cos\phi = 1 - \frac{\lambda'_{\max} - \lambda}{\lambda_C} = 1 - \frac{\lambda'_{\max}}{\lambda_C} + \frac{(hc/E)}{\lambda_C} = \frac{1}{2} + \frac{mc^2}{E}.$$

This expression shows that for $E \gg mc^2$, $\cos\phi = \frac{1}{2}$ and so $\phi = 60°$, but it also shows that, because $\cos\phi$ must always be less than 1, for pair production at *any* angle, E must be greater than $2mc^2$, which we know to be the case.

2-43: (a) The most direct way to get this result is to use Equation (2.26) with $I_0/I = 2$, so that

$$x_{1/2} = \frac{\ln(2)}{\mu} = \frac{0.693}{\mu}.$$

(b) Similarly, with $I_0/I = 10$,

$$x_{1/10} = \frac{\ln(10)}{\mu} = \frac{2.30}{\mu}.$$

2-45: From either Equation (2.26) or Problem 2-43 above,

$$x_{1/2} = \frac{\ln(2)}{\mu} = \frac{0.693}{78 \text{ m}^{-1}} = 8.9 \times 10^{-3} \text{ m} = 8.9 \text{ mm}.$$

2-47: Rather than calculating the actual intensity ratios, Equation (2.26) indicates that the ratios will be the same when the distances in water and lead are related by

$$(\mu_{H_2O})(x_{H_2O}) = (\mu_{Pb})(x_{Pb}), \quad \text{or}$$

$$x_{H_2O} = x_{Pb}\frac{\mu_{Pb}}{\mu_{H_2O}} = (10 \times 10^{-3}\text{ m})\frac{(52\text{ m}^{-1})}{(4.9\text{ m}^{-1})} = 0.106\text{ m},$$

or 11 cm to two significant figures.

2-49: Either a direct application of Equation (2.26) or use of the result of Problem 2-43 gives

$$x_{1/2} = \frac{\ln(2)}{4.7 \times 10^4\text{ m}^{-1}} = 1.47 \times 10^{-5}\text{ m},$$

which is 0.015 mm to two significant figures.

2-51: In Equation (2.29), the ratio

$$\frac{GM}{c^2 R} = \frac{(6.67 \times 10^{-11}\text{ N·m}^2/\text{kg}^2)(2.0 \times 10^{30}\text{ kg})}{(2.998 \times 10^8\text{ m/s})^2(7.0 \times 10^8\text{ m})} = 2.12 \times 10^{-6}$$

(keeping an extra significant figure) is so small that for an "approximate" red shift, the ratio $\Delta\lambda/\lambda$ will be the same as $\Delta\nu/\nu$, and

$$\Delta\lambda = \lambda\frac{GM}{c^2 R} = (500 \times 10^{-9}\text{ m})(2.12 \times 10^{-6}) = 1.06 \times 10^{-12}\text{ m} = 1.06\text{ pm}.$$

2-53: (a) The most convenient way to do this problem, for computational purposes, is to realize that the nucleus will be moving nonrelativistically after the emission of the photon, and that the energy of the photon will be very close to $E_\infty = 14.4$ keV, the energy that the photon would have if the nucleus had been infinitely massive. So, if the photon has an energy E, the recoil momentum of the nucleus is E/c, and its kinetic energy is $\frac{p^2}{2M} = \frac{E^2}{2Mc^2}$, where M is the rest mass of the nucleus. Then, conservation of energy implies

$$\frac{E^2}{2Mc^2} + E = E_\infty.$$

This is a quadratic in E, and solution might be attempted by standard methods, but to find the *change* in energy due to the finite mass of the nucleus, and recognizing that E will be very close to E_∞, the above relation may be expressed as

$$E_\infty - E = \frac{E^2}{2\,Mc^2} \approx \frac{E_\infty^2}{2\,Mc^2}$$
$$= \frac{(14.4\text{ keV})^2 \left(1.602 \times 10^{-16}\text{ Jk/eV}\right)}{2\,(9.5 \times 10^{-26}\text{ kg})\,(2.998 \times 10^8\text{ m/s})^2}$$
$$= 1.9 \times 10^{-6}\text{ keV} = 1.9 \times 10^{-3}\text{ eV}.$$

If the approximation $E \approx E_\infty$ is not made, the resulting quadratic is

$$E^2 + 2\,Mc^2\,E - 2\,Mc^2\,E_\infty = 0,$$

which is solved for

$$E = Mc^2 \left[\sqrt{1 + \left(2\,\frac{E_\infty}{Mc^2}\right)} - 1\right].$$

However, the dimensionless quantity $E_\infty/(Mc^2)$ is so small that standard calculators are not able to determine the difference between E and E_∞. The square root must be expanded, using $(1+x)^{1/2} \approx 1 + (x/2) - (x^2/8)$, and two terms must be kept to find the *difference* between E and E_∞. This approximation gives the previous result.

It so happens that a relativistic treatment of the recoiling nucleus gives the same numerical result, but without intermediate approximations or solution of a quadratic equation. The relativistic form expressing conservation of energy is, with $pc = E$ and before,

$$\sqrt{E^2 + (Mc^2)^2} + E = Mc^2 + E_\infty, \quad \text{or} \quad \sqrt{E^2 + (Mc^2)^2} = Mc^2 + E_\infty - E.$$

Squaring both sides, canceling E^2 and $(Mc^2)^2$, and then solving for E,

$$E = \frac{E_\infty^2 + 2\,Mc^2\,E_\infty}{2\,(Mc^2 + E_\infty)} = E_\infty \left(\frac{1 + (E_\infty/(2\,Mc^2))}{1 + (E_\infty/(Mc^2))}\right).$$

From this form,

$$E_\infty - E = \left(\frac{E_\infty^2}{2\,Mc^2}\right)\frac{1}{1 + (E_\infty/(Mc^2))},$$

giving the same result.

(b) For this situation, the above result applies, but the nonrelativistic approximation is by far the easiest for calculation;

$$E_\infty - E = \frac{E_\infty^2}{2\,Mc^2} = \frac{(14.4 \times 10^3\text{ eV})^2 \left(1.602 \times 10^{-19}\text{ J/eV}\right)}{2\,(1.0 \times 10^{-3}\text{ kg})\,(2.998 \times 10^8\text{ m/s})^2} = 1.8 \times 10^{-25}\text{ eV}.$$

(c) The original frequency is

$$\nu = \frac{E_\infty}{h} = \frac{14.4 \times 10^3 \text{ eV}}{4.136 \times 10^{-15} \text{ eV} \cdot \text{s}} = 3.48 \times 10^{18} \text{ Hz}.$$

From Equation (2.28), the change in frequency is

$$\Delta\nu = \nu' - \nu = \left(\frac{gH}{c^2}\right)\nu = \frac{(9.8 \text{ m/s}^2)(20 \text{ m})}{(2.998 \times 10^8 \text{ m/s})^2}(3.48 \times 10^{18} \text{ Hz}) = 7.6 \text{ Hz}.$$

2-55: (a) To leave the body of mass M permanently, the body of mass m must have enough kinetic energy so that there is no radius at which its energy is positive. That is, its total energy must be non-negative. The escape velocity v_e is the speed (for a given radius, and assuming $M \gg m$) that the body of mass m would have for a total energy of zero;

$$\frac{1}{2}mv_e^2 - \frac{GMm}{R} = 0, \quad \text{or} \quad v_e = \sqrt{\frac{2GM}{R}}.$$

(b) Solving the above expression for R in terms of v_e,

$$R = \frac{2GM}{v_e^2},$$

and if $v_e = c$, Equation (2.30) is obtained.

Chapter 3 - Wave Properties of Particles

3-1: From Equation (3.1), any particle's wavelength is determined by its momentum, and hence particles with the same wavelength have the same momenta. With a common momentum p, the photon's energy is pc, and the particle's energy is $\sqrt{(pc)^2 + (mc^2)^2}$, which is necessarily greater than pc for a massive particle. The particle's kinetic energy is

$$\text{KE} = E - mc^2 = \sqrt{(pc)^2 + (mc^2)^2} - mc^2.$$

For low values of p ($p \ll mc$ for a nonrelativistic massive particle), the kinetic energy is $\text{KE} \approx \dfrac{p^2}{2m}$, which is necessarily less than pc. For a realtivistic massive particle, $\text{KE} \approx pc - mc^2$, and KE is less than the photon energy. The kinetic energy of a massive particle will always be less than pc, as can be seen by using $E = \text{KE} + mc^2$, squaring, and subtracting from $E^2 = (pc)^2 + (mc^2)^2$ to obtain

$$(pc)^2 - \text{KE}^2 = 2\,\text{KE}\,mc^2.$$

3-3: For this nonrelativistic case,

$$\lambda = \frac{h}{mv} = \frac{(6.626 \times 10^{-34}\text{ J·s})}{(1.0 \times 10^{-6}\text{ kg})(20\text{ m/s})} = 3.3 \times 10^{-29}\text{ m};$$

quantum effects certainly would not be noticed for such an object.

3-5: Becasue the de Broglie wavelength depends only on the electron's momentum, the percentage error in the wavelength will be the same as the percentage error in the reciprocal of the momentum, with the nonrelativistic calculation giving the higher wavelength due to a lower calculated momentum. The nonrelativistic momentum is

$$p_{\text{nr}} = \sqrt{2m\,\text{KE}} = \sqrt{2\,(9.1095 \times 10^{-31}\text{ kg})(100 \times 10^3\text{ eV})(1.602 \times 10^{-19}\text{ J/eV})}$$
$$= 1.708 \times 10^{-22}\text{ kg·m/s},$$

and the nonrelativistic momentum is

$$p_{\text{r}} = \frac{1}{c}\sqrt{(\text{KE} + mc^2)^2 - (mc^2)^2} = \sqrt{(0.100 + 0.511)^2 - (0.511)^2}\text{ MeV}/c$$
$$= 1.790 \times 10^{-22}\text{ kg·m/s},$$

keeping extra figures in the intermediate calculations. The percentage error in the computed de Broglie wavelength is then

$$\frac{(h/p_{\text{nr}}) - (h/p_{\text{r}})}{(h/p_{\text{r}})} = \frac{p_{\text{r}} - p_{\text{nr}}}{p_{\text{nr}}} = \frac{1.790 - 1.708}{1.708} = 4.8\%.$$

3-7: A nonrelativistic calculation gives

$$\text{KE} = \frac{p^2}{2m} = \frac{(hc/\lambda)^2}{2mc^2} = \frac{(hc)^2}{2(mc^2)\lambda^2}$$

$$= \frac{\left(1.240 \times 10^{-6}\ \text{eV}\cdot\text{m}\right)^2}{2\left(939.6 \times 10^6\ \text{eV}\right)\left(0.282 \times 10^{-9}\ \text{m}\right)^2} = 0.00103\ \text{eV}.$$

(Note that in the above caculation, multiplication of numerator and denominator by c^2 and use of the product hc in terms of electronvolts avoided further unit conversion.) This energy is much less than the neutron's rest energy, and so the nonrelativistic calculation is completely valid.

3-9: A nonrelativistic calculation gives

$$\text{KE} = \frac{p^2}{2m} = \frac{(hc/\lambda)^2}{2mc^2} = \frac{(hc)^2}{2(mc^2)\lambda^2}$$

$$= \frac{\left(1.240 \times 10^{-6}\ \text{eV}\cdot\text{m}\right)^2}{2\left(511 \times 10^3\ \text{eV}\right)\left(550 \times 10^{-9}\ \text{m}\right)^2} = 5.0 \times 10^{-6}\ \text{eV},$$

so the electron would have to be accelerated through a potential difference of $5.0 \times 10^{-6}\ \text{V} = 5.0\ \mu\text{V}$. Note that the kinetic energy is very small compared to the electron rest energy, so the nonrelativistic calculation is valid. (In the above caculation, multiplication of numerator and denominator by c^2 and use of the product hc in terms of electronvolts avoided further unit conversion.)

3-11: If $E^2 = (pc)^2 + \left(mc^2\right)^2 \gg \left(mc^2\right)^2$, then $pc \gg mc^2$ and $E \approx pc$. For a photon with the same energy, $E = pc$, so the momentum of such a particle would be nearly the same as a photon with the same energy, and so the de Broglie wavelengths would be the same.

3-13: For massive particles of the same speed, relativistic or nonrelativistic, the momentum will be proportional to the mass, and so the de Broglie wavelength will be inversely proportional to the mass; the electron will have the longer wavelength by a factor of $(m_p/m_e) = 1838$. From Equation (3.3) the particles have the same phase velocity and from Equation (3.16) they have the same group velocity.

3-15: Suppose that the phase velocity is independent of wavelength, and hence independent of the wave number k; then, from Equation (3.3), the phase velocity $v_p = (\omega/k) = u$, a constant. It follows that because $\omega = u\,k$,

$$v_g = \frac{d\omega}{dk} = u = v_p.$$

3-17: The phase velocity may be expressed in terms of the wave number $k = 2\pi/\lambda$ as

$$v_p = \frac{\omega}{k} = \sqrt{\frac{g}{k}}, \quad \text{or} \quad \omega = \sqrt{gk} \quad \text{or} \quad \omega^2 = g\,k.$$

Finding the group velocity by differentiating $\omega(k)$ with respect to k,

$$v_g = \frac{d\omega}{dk} = \frac{1}{2}\sqrt{g}\,\frac{1}{\sqrt{k}} = \frac{1}{2}\sqrt{\frac{g}{k}} = \frac{1}{2}\frac{\omega}{k} = \frac{1}{2}v_g.$$

Using implicit differentiation in the formula for $\omega^2(k)$,

$$2\omega\frac{d\omega}{dk} = 2\omega v_g = g,$$

So that
$$v_g = \frac{g}{2\omega} = \frac{g\,k}{2\omega k} = \frac{\omega^2}{2\omega k} = \frac{\omega}{2k} = \frac{1}{2}v_p,$$

the same result. For those more comfortable with calculus, the dispersion relation may be expressed as

$$2\ln(\omega) = \ln(k) + \ln(g),$$

from which $2\dfrac{d\omega}{\omega} = \dfrac{dk}{k}$, and $v_g = \dfrac{1}{2}\dfrac{\omega}{k} = \dfrac{1}{2}v_p$.

3-19: For a kinetic energy of 500 keV,

$$\gamma = \frac{1}{\sqrt{1 - \dfrac{v^2}{c^2}}} = \frac{KE + mc^2}{mc^2} = \frac{500 + 511}{511} = 1.978.$$

Solving for v,

$$v = c\sqrt{1 - \frac{1}{\gamma}} = \sqrt{1 - \left(\frac{1}{1.978}\right)^2} = 0.863\,c,$$

and from Equation (3.16), $v_g = v = 0.863\,c$. The phase velocity is then $v_p = c^2/v_g = 1.16\,c$.

3-21: (a) Two equivalent methods will be presented here. Both will assume the validity of Equation (3.16), in that $v_g = v$.

First: Express the wavelength λ in terms of v_g,

$$\lambda = \frac{h}{p} = \frac{h}{m v_g \gamma} = \frac{h}{m v_g}\sqrt{1 - \frac{v_g^2}{c^2}}.$$

Multiplying by $m v_g$, squaring and solving for v_g^2 gives

$$v_g^2 = \frac{h^2}{(\lambda m)^2 + (h^2/c^2)} = c^2\left[1 + \left(\frac{m\lambda c}{h}\right)^2\right]^{-1}.$$

Taking the square root and using Equation (3.3), $v_p = c^2/v_g$, gives the desired result.

Second: Consider the particle energy in terms of $v_p = c^2/v_g$;

$$E^2 = (pc)^2 + (mc^2)^2$$

$$\gamma^2 (mc^2)^2 = \frac{(mc^2)^2}{1 - \frac{c^2}{v_p^2}} = \left(\frac{hc}{\lambda}\right)^2 + (mc^2)^2.$$

Dividing by $(mc^2)^2$ leads to

$$1 - \frac{c^2}{v_p^2} = \frac{1}{1 + h^2/(mc\lambda)^2}, \qquad \text{so that}$$

$$\frac{c^2}{v_p^2} = 1 - \frac{1}{1 + h^2/(mc\lambda)^2} = \frac{h^2(mc\lambda)^2}{h^2(mc\lambda)^2 + 1} = \frac{1}{1 + (mc\lambda)^2/h^2},$$

which is an equivalent statement of the desired result.

It should be noted that in the first method presented above could be used to find λ in terms of v_p directly, and in the second method the energy could be found in terms of v_g. The final result is, or course, the same.

(b) Using the result of part (a),

$$v_p = c\sqrt{1 + \left(\frac{(9.1095 \times 10^{-31}\text{ kg})(2.998 \times 10^8\text{ m/s})(1.00 \times 10^{-13}\text{ m})}{6.626 \times 10^{-34}\text{ J}\cdot\text{s}}\right)^2}$$
$$= 1.00085\, c,$$

and $v_g = c^2/v_g = 0.99915\, c$.

Wave Properties of Particles 29

For a calculational shortcut, write the result of part (a) as

$$v_p = c\sqrt{1 + \left(\frac{mc^2 \lambda}{hc}\right)^2}$$

$$= c\sqrt{1 + \left(\frac{(511 \times 10^3 \text{ eV})(1.00 \times 10^{-13} \text{ m})}{(1.240 \times 10^{-6} \text{ eV·m})}\right)^2} = 1.00085\, c.$$

In both of the above answers, the statement that the de Broglie wavelength is "exactly" 10^{-13} m means that the answers can be given to any desired precision.

3-23: Increasing the electron energy increases the electron's momentum, and hence decreases the electron's de Broglie wavelength. From Equation (2.13), a smaller de Broglie wavelength results in a smaller scattering angle.

3-25: (a) For the given energies, a nonrelativistic calculation is sufficient;

$$v = \sqrt{\frac{2\,\text{KE}}{m}} = \sqrt{\frac{2\,(54\text{ eV})(1.602 \times 10^{-19}\text{ J/eV})}{(9.1095 \times 10^{-31}\text{ kg})}} = 4.36 \text{ m/s}$$

outside the crystal, and (from a similar calculation, with KE = 80 eV), $v = 5.30 \times 10^6$ m/s inside the crystal (keeping an extra significant figure in both calculations).

(b) With the speeds found in part (a), the de Broglie wavelengths are found from

$$\lambda = \frac{h}{p} = \frac{h}{m\,v} = \frac{(6.626 \times 10^{-34}\text{ J·s})}{(9.1095 \times 10^{-31}\text{ kg})(4.36 \times 10^6\text{ m/s})} = 1.67 \times 10^{-10}\text{ m},$$

or 0.167 nm outside the crystal, with a similar calculation giving 0.137 nm inside the crystal.

3-27: From Equation (3.18),

$$E_n = n^2 \frac{h^2}{8\,m\,L^2} = n^2 \frac{(6.626 \times 10^{-34}\text{ J·s})^2}{8\,(1.675 \times 10^{-27}\text{ kg})(1.00 \times 10^{-14}\text{ m})^2}$$

$$= n^2\, 3.28 \times 10^{-13}\text{ J} = n^2\, 20.5\text{ MeV}.$$

The minimum energy, corresponding to $n = 1$, is 20.5 MeV.

3-29: The first excited state corresponds to $n = 2$ in Equation (3.18). Solving for the width L,

$$L = n\sqrt{\frac{h^2}{8\,m\,E_2}} = 2\sqrt{\frac{(6.626 \times 10^{-34}\text{ J·s})^2}{8\,(1.673 \times 10^{-27}\text{ kg})(400 \times 10^3\text{ eV})(1.602 \times 10^{-19}\text{ J/eV})}}$$

$$= 4.53 \times 10^{-14}\text{ m} = 45.3\text{ fm}.$$

3-31: Each atom in a solid is limited to a certain definite region of space – otherwise the assembly of atoms would not be a solid. The uncertainty in position of each atom is therefore finite, and its momentum and hence enrgy cannot be zero. The position of an ideal-gas molecule is not restricted, so the uncertainty in its position is effectively infinite and its momentum and hence enrgy can be zero.

3-33: The percentage uncertainty in the electron's momentum will be at least

$$\frac{\Delta p}{p} = \frac{h}{4\pi \Delta x\, p} = \frac{h}{4\pi \Delta x\, \sqrt{2m\,\text{KE}}} = \frac{hc}{4\pi \Delta x\, \sqrt{2\,(mc^2)\,\text{KE}}}$$

$$= \frac{(1.240 \times 10^{-6}\ \text{eV}\cdot\text{m})}{4\pi\,(1.00 \times 10^{-10}\ \text{m})\sqrt{2\,(511 \times 10^3\ \text{eV})(1.00 \times 10^3\ \text{eV})}}$$

$$= 3.1 \times 10^{-2} = 3.1\%.$$

Note that in the above calculation, conversion of the mass of the electron into its energy equivalent in electronvolts is purely optional; converting the kinetic energy into joules and using $h = 6.626 \times 10^{-34}$ J·s will of course give the same percentage uncertainty.

3-35: The proton will need to move a minimum distance

$$v\,\Delta t \geq v\,\frac{h}{4\pi\,\Delta E},$$

where v can be taken to be

$$v = \sqrt{\frac{2\,\text{KE}}{m}} = \sqrt{\frac{2\,\Delta E}{m}}, \quad \text{so that}$$

$$v\,\Delta t = \sqrt{\frac{2\,\text{KE}}{m}}\,\frac{h}{4\pi\,\Delta E} = \frac{h}{2\pi\,\sqrt{2m\,\text{KE}}} = \frac{hc}{2\pi\,\sqrt{2\,(mc^2)\,\text{KE}}}$$

$$= \frac{1.240 \times 10^{-6}\ \text{eV}\cdot\text{m}}{2\pi\sqrt{2\,(938.28 \times 10^6\ \text{MeV})(1.00 \times 10^3\ \text{eV})}}$$

$$= 1.44 \times 10^{-13}\ \text{m} = 0.144\ \text{pm}.$$

(See note to the solution to Problem 3-33 above).

The result for the product $v\,\Delta t$ may be recognized as $v\,\Delta t \geq \dfrac{h}{2\pi p}$; this is not inconsistent with Equation (3.21), $\Delta x\,\Delta p \geq \dfrac{h}{4\pi}$. In the current problem, ΔE was taken to be the (maximum) kinetic energy of the proton. In such a situation,

$$\Delta E = \frac{\Delta(p^2)}{m} = 2\,\frac{p}{m}\,\Delta p = 2\,v\,\Delta p,$$

which is consistent with the previous result.

Wave Properties of Particles 31

3-37: (a) The length of each group is

$$c\,\Delta t = (2.998 \times 10^8 \text{ m/s})(8.00 \times 10^{-8} \text{ s}) = 24 \text{ m}.$$

The number of waves in each group is the pulse duration divided by the wave period, which is the pulse duration multiplied by the frequency,

$$(8.00 \times 10^{-8} \text{ s})(4900 \times 10^6 \text{ Hz}) = 752 \text{ waves}.$$

(b) The bandwidth is the reciprocal of the pulse duration,

$$(8.00 \times 10^{-8} \text{ s})^{-1} = 12.5 \text{ MHz}.$$

3-39: To use the uncertainty principle, make the identification of p with Δp and x with Δx, so that $p = h/(4\pi x)$, and

$$E = E(x) = \left(\frac{h^2}{8\pi^2 m}\right)\frac{1}{x^2} + \left(\frac{C}{2}\right)x^2.$$

Differentiating with respect to x and setting $\dfrac{d}{dx}E = 0$,

$$-\left(\frac{h^2}{4\pi^2 m}\right)\frac{1}{x^3} + Cx = 0,$$

which is solved for

$$x^2 = \frac{h}{2\pi\sqrt{mC}}.$$

Substution of this value into $E(x)$ gives

$$E_{\min} = \left(\frac{h^2}{8\pi^2 m}\right)\left(\frac{2\pi\sqrt{mC}}{h}\right) + \left(\frac{C}{2}\right)\left(\frac{h}{2\pi\sqrt{mC}}\right) = \frac{h}{2\pi}\sqrt{\frac{C}{m}} = \frac{h\nu}{2}.$$

Chapter 4 - Atomic Structure

4-1: The fact that most particles pass through undeflected means that there is not much to deflect these particles; most of the volume of an atom is empty space, and gases and metals are overall electrically neutral.

4-3: For a "closest approach", the incident proton must be directed "head-on" to the nucleus, with no angular momentum with respect to the nucleus (an "impact parameter" of zero; see the Appendix to Chapter 4). In this case, at the point of closest approach the proton will have no kinetic energy, and so the potential energy at closest approach will be the initial kinetic energy, taking the potential energy to be zero in the limit of very large separation. Equating these energies,

$$\text{KE}_{\text{initial}} = \frac{Z e^2}{4\pi \epsilon_0 r_{\text{min}}}, \quad \text{or}$$

$$r_{\text{min}} = \left(\frac{1}{4\pi \epsilon_0}\right) \frac{Z e^2}{\text{KE}_{\text{initial}}} = (8.988 \times 10^9 \text{ N·m}^2/\text{C}^2) \frac{(79)\left(1.602 \times 10^{-19} \text{ C}\right)^2}{(1.602 \times 10^{-13} \text{ J})}$$
$$= 1.14 \times 10^{-13} \text{ m}.$$

4-5: The wavelengths in the Brackett series are given in Equation (4.9); the shortest wavelength (highest energy) corresponds to the largest value of n. For $n \to \infty$,

$$\lambda \to \frac{16}{R} = \frac{16}{1.097 \times 10^7 \text{ m}^{-1}} = 1.46 \times 10^{-6} \text{ m} = 1.46 \text{ } \mu\text{m}.$$

4-7: While the kinetic energy of any particle is positive, the potential energy of any pair of particles that are mutually attracted is negative. For the system to be bound, the total energy, the sum of the positive kinetic energy and the total negative potential energy, must be negative. For a classical particle subject to an inverse-square attractive force (such as two oppositely charged particles or two uniform spheres subject to gravitational attraction) in a circular orbit, the potential energy is twice the negative of the kinetic energy.

Atomic Structure 33

4-9: (a) The velocity v_1 is given by Equation (4.4), with $r = r_1 = a_0$. Combining to find v_1^2,

$$v_1^2 = \frac{e^2}{4\pi \epsilon_0 \, m \, a_0} = \frac{e^2}{4\pi \epsilon_0 \, m \left(\dfrac{h^2 \epsilon_0}{\pi \, m \, e^2}\right)} = \frac{e^4}{4\, \epsilon_0^2 \, h^2}, \quad \text{so}$$

$$\frac{v_1}{c} = \frac{e^2}{2\,\epsilon_0\, h}\frac{1}{c} = \alpha.$$

(b) From the above,

$$\alpha = \frac{\left(1.602 \times 10^{-19}\ \text{C}\right)^2}{2\,(8.854 \times 10^{-12}\ \text{C}^2/(\text{N·m}^2))\,(6.626 \times 10^{-34}\ \text{J·s})\,(2.998 \times 10^8\ \text{m/s})}$$
$$= 7.296 \times 10^{-3},$$

so that $\dfrac{1}{\alpha} = 137.1$ to four significant figures.

A close check of the units is worthwhile; treating the units as algebraic quantities, the units as given in the above calculation are

$$\frac{[\text{C}^2]}{\dfrac{[\text{C}^2]}{[\text{N}]\,[\text{m}^2]}\,[\text{J}]\,[\text{s}]\,\dfrac{[\text{m}]}{[\text{s}]}} = \frac{[\text{N·m}]}{[\text{J}]} = 1.$$

Thus, α is a dimensionless quantity, and will have the same numerical value in any system of units.

The most accurate (November, 2001) value of $1/\alpha$ is

$$\frac{1}{\alpha} = 137.03599976,$$

accurate to better than 4 parts per *billion*. For the most accurately known values of this or other physical constants, see, for instance, the **Particle Data Group** tables of **Constants, Units, Atomic and Nuclear Properties**, available at
http://pdg.lbl.gov/2001/contents_sports.html

(c) Using the above expression for α and Equation (4.13) with $n = 1$ for a_0,

$$\alpha\, a_0 = \frac{e^2}{2\,\epsilon_0\, h\, c}\frac{h^2 \epsilon_0}{\pi\, m\, e^2} = \frac{1}{2\pi}\frac{h}{m\,c} = \frac{\lambda_C}{2\pi},$$

where the Compton wavelength λ_C is given by Equation (2.22).

4-11: With the mass, orbital speed and orbital radius of the earth known, the earth's orbital angular momentum is known, and the quantum number that would characterize the earth's orbit about the sun would be this angular momentum divided by $\hbar$;

$$n = \frac{L}{\hbar} = \frac{mvR}{\hbar} = \frac{(6.0 \times 10^{24} \text{ kg})(3.0 \times 10^4 \text{ m/s})(1.5 \times 10^{11} \text{ m})}{(1.055 \times 10^{-34} \text{ J·s})} = 2.6 \times 10^{74}.$$

(The number of significant figures not of concern.)

4-13: The uncertainty in position of an electron confined to such a region is, from Equation (3.22), $\Delta p \geq \dfrac{\hbar}{2a_0}$, while the magnitude of the linear momentum of an electron in the first Bohr orbit is

$$p = \frac{h}{\lambda} = \frac{h}{2\pi a_0} = \frac{\hbar}{a_0};$$

the value of Δp found from Equation (3.13) is half of this momentum.

4-15: The Doppler effect shifts the frequencies of the emitted light to both higher and lower frequencies to produce wider lines than atoms at rest would give rise to.

4-17: It must assumed that the initial electrostatic potential energy is negligible, so that the final energy of the hydrogen atom is $E_1 = -13.6$ eV. The energy of the photon emitted is then $-E_1$, and the wavelength is

$$\lambda = \frac{hc}{(-E_1)} = \frac{1.240 \times 10^{-6} \text{ eV·m}}{13.6 \text{ eV}} = 9.12 \times 10^{-8} \text{ m} = 91.2 \text{ nm},$$

in the ultraviolet part of the spectrum (see, for instance, the back endpapers of the text).

4-19: From either Equation (4.7) with $n = 10$ or Equation (4.18) with $n_f = 1$ and $n_i = 10$,

$$\lambda = \frac{100}{99}\frac{1}{R} = \frac{100}{99}\frac{1}{1.097 \times 10^7 \text{ m}^{-1}} = 9.21 \times 10^{-8} \text{ m} = 92.1 \text{ nm},$$

which is in the ultraviolet part of the spectrum (see, for instance, the back endpapers of the text).

Atomic Structure 35

4-21: The electrons' energy must be at least the difference between the $n = 1$ and $n = 3$ levels,

$$\Delta E = E_3 - E_1 = -E_1 \left(1 - \frac{1}{9}\right) = (13.6 \text{ eV}) \frac{8}{9} = 12.1 \text{ eV}$$

(this assumes that few or none of the hydrogen atoms had electrons in the $n = 2$ level). A potential difference of 12.1 eV is necessary to accelerate the electrons to this energy.

4-23: The energy needed to ionize hydrogen will be the energy needed to raise the energy from the ground state to the first excited state plus the energy needed to ionize an atom in the second excited state; these are the energies that correspond to the longest wavelength (least energetic photon) in the Lyman series and the shortest wavelength (most energetic photon) in the Balmer series. The energies are proportional to the reciprocals of the wavelengths, and so the wavelength of the photon needed to ionize hydrogen is

$$\lambda = \left(\frac{1}{\lambda_{2 \to 1}} + \frac{1}{\lambda_{\infty \to 2}}\right)^{-1} = \left(\frac{1}{121.5 \text{ nm}} + \frac{1}{364.6 \text{ nm}}\right)^{-1} = 91.13 \text{ nm}.$$

As a check, note that this wavelength is R^{-1}.

4-25: (a) From Equation (4.7) with $n = n_i$,

$$\frac{1}{\lambda} = R\left(1 - \frac{1}{n_i^2}\right),$$

which is solved for

$$n_i = \left(1 - \frac{1}{\lambda R}\right)^{-1/2} = \sqrt{\frac{\lambda R}{\lambda R - 1}}.$$

(b) Either of the above forms gives n very close (four places) to 3; specifically, with the product $\lambda R = (102.55 \times 10^{-9} \text{ m})(1.097 \times 10^7 \text{ m}^{-1}) = 1.125$ rounded to four places as $\frac{9}{8}$, $n = 3$ exactly.

4-27: (a) A relativistic calculation would necessarily involve the change in mass of the atom due to the change in energy of the system. The fact that this mass change is too small to measure (that is, the change is measured indirectly by measuring the energies of the emitted photons) means that a nonrelativistic calculation should suffice. In this situation, the kinetic energy of the recoiling atom is

$$\text{KE} = \frac{p^2}{2M} = \frac{(h\nu/c)^2}{2M},$$

where ν is the frequency of the emitted photon and $p = h/\lambda = h\nu/c$ is the magnitude of the momentum of both the photon and the recoiling atom. Equation (4.16) is then

$$E_i - E_f = h\nu + \text{KE} = h\nu + \frac{(h\nu)^2}{2Mc^2} = h\nu\left(1 + \frac{h\nu}{2Mc^2}\right).$$

This result is equivalent to that of Problem 2-53, where $h\nu = E_\infty$ and the term $p^2/(2M)$ corresponds to $E_\infty - E$ in that problem. As in Problem 2-53, a relativistic calculation is managable; the result would be

$$E_f - E_i = h\nu\left(1 + \frac{1}{2}\left(1 + \frac{Mc^2}{h\nu}\right)^{-1}\right),$$

a form not often useful; see part (b).

(b) As indicated above and in the problem statement, a nonrelativistic calculation is sufficient. As in part (a),

$$\text{KE} = \frac{p^2}{2M} = \frac{(\Delta E/c)^2}{2M}, \quad \text{and}$$

$$\frac{\text{KE}}{\Delta E} = \frac{\Delta E}{2Mc^2} = \frac{1.9 \text{ eV}}{2(939 \times 10^6 \text{ eV})} = 1.01 \times 10^{-9},$$

or 1.0×10^{-9} to two significant figures. In the above, the rest energy of the hydrogen atom is from the front endpapers.

4-29: There are many equivalent algebraic methods that may be used to derive Equation (4.19), and that result will be cited here;

$$f_n = -\frac{2E_1}{h}\frac{1}{n^3}.$$

The frequency ν of the photon emitted in going from the level $n+1$ to the level n is obtained from Equation (4.17) with $n_i = n+1$ and $n_f = n$;

$$\nu = \frac{\Delta E}{h}\left[\frac{1}{(n+1)^2} - \frac{1}{n^2}\right] = -\frac{2E_1}{h}\left[\frac{n+\frac{1}{2}}{n^2(n+1)^2}\right].$$

This can be seen to be equivalent to the expression for ν in terms of n and p that was found in the derivation of Equation (4.20), but with n replaced by $n+1$ and $p = 1$. Note that in this form, ν is positive because E_1 is negative.

Atomic Structure 37

From this expression,

$$\nu = -\frac{2E_1}{h n^3}\left[\frac{n^2+\frac{1}{2}n}{n^2+2n+1}\right] = f_n\left[\frac{n^2+\frac{1}{2}n}{n^2+2n+1}\right] < f_n,$$

as the term in brackets is less than 1. Similarly,

$$\nu = -\frac{2E_1}{h(n+1)^3}\left[\frac{(n+\frac{1}{2})(n+1)}{n^2}\right] = f_{n+1}\left[\frac{(n+\frac{1}{2})(n+1)}{n^2}\right] > f_{n+1},$$

as the term in brackets is greater than 1.

4-31: For a muonic atom, the Rydberg constant is multiplied by the ratio of the reduced masses of the muoninc atom and the hydrogen atom, $R' = R(m'/m_e) = 186R$, as in Example 4.7; from Equation (4.7),

$$\lambda = \frac{4/3}{R'} = \frac{4/3}{186(1.097\times 10^7\text{ m}^{-1})} = 6.53\times 10^{-10}\text{ m} = 0.653\text{ nm},$$

in the x-ray range.

4-33: The H_α lines, corresponding to $n=3$ in Equation (4.6), have wavelengths of $\lambda = (36/5)(1/R)$. For a tritium atom, the wavelength would be $\lambda_T = (36/5)(1/R_T)$, where R_T is the Rydberg constant evaluated with the reduced mass of the tritium atom replacing the reduced mass of the hydrogen atom. The difference between the wavelengths would then be

$$\Delta\lambda = \lambda - \lambda_T = \lambda\left[1-\frac{\lambda_T}{\lambda}\right] = \lambda\left[1-\frac{R}{R_T}\right].$$

The values of R and R_T are proportional to the respective reduced masses, and their ratio is

$$\frac{R}{R_T} = \frac{m_e m_H/(m_e+m_H)}{m_e m_T/(m_e+m_T)} = \frac{m_H(m_e+m_T)}{m_T(m_e+m_H)}.$$

Using this in the above expression for $\Delta\lambda$,

$$\Delta\lambda = \lambda\left[\frac{m_e(m_T-m_H)}{m_e(m_e+m_H)}\right] \approx \lambda\frac{2m_e}{3m_H},$$

where the approximations $m_e + m_H \approx m_H$ and $m_T \approx 3m_H$ have been used. Inserting numerical values,

$$\Delta\lambda = \frac{(36/5)}{(1.097\times 10^7\text{ m}^{-1})}\frac{2(9.1095\times 10^{-31}\text{ kg})}{3(1.6736\times 10^{-27}\text{ kg})} = 2.38\times 10^{-10}\text{ m} = 0.238\text{ nm}.$$

4-35: (a) The steps leading to Equation (4.15) are repeated, with Ze^2 instead of e^2 and Z^2e^4 instead of e^4, giving

$$E_n = -\frac{m' Z^2 e^4}{8\pi \epsilon_0^2 h^2} \frac{1}{n^2},$$

where the reduced mass m' will depend on the mass of the nucleus.

(b) A plot of the energy levels is given below. The scale is close, but not exact, and of course there are many more levels corresponding to higher n. In the approximation that the reduced masses are the same, for He$^+$, with $Z = 2$, the $n = 2$ level is the same as the $n = 1$ level for Hydrogen, and the $n = 4$ level is the same as the $n = 2$ level for hydrogen.

The energy levels for H and He$^+$:

	H$^+$		H	
				$n \to \infty$
-3.4 eV	_____	$n=4$	_____	$n=2$
-6.0 eV	_____	$n=3$		
-13.6 eV	_____	$n=2$	_____	$n=1$
-54.4 eV	_____	$n=1$		

(c) When the electron joins the Helium nucleus, the electron-nucleus system loses energy; the emitted photon will have lost energy $\Delta E = 4(-13.6 \text{ eV}) = -54.4$ eV, where the result of part (a) has been used. The emitted photon's wavelength is

$$\lambda = \frac{hc}{-\Delta E} = \frac{1.240 \times 10^{-6} \text{ eV·m}}{54.4 \text{ eV}} = 2.28 \times 10^{-8} \text{ m} = 22.8 \text{ nm}.$$

4-37: The minimum number of Cr^{3+} ions will be the minimum number of photons, which is the total energy of the pulse divided by the energy of each photon,

$$\frac{E}{hc/\lambda} = \frac{E\lambda}{hc} = \frac{(1.00 \text{ J})(694 \times 10^{-9} \text{ m})}{(6.626 \times 10^{-34} \text{ J·s})(2.998 \times 10^8 \text{ m/s})} = 3.49 \times 10^{18} \text{ ions}.$$

Atomic Structure 39

4-39: Small angles correspond to particles that are not scattered much at all, and the structure of the atom does not affect these particles. To these nonpenetrating particles, the nucleus is either partially or completely screened by the atom's electron cloud, and the scattering analysis, based on a pointlike positively charged nucleus, is not applicable.

4-41: From Equation (4.29), using the value for $\frac{1}{4\pi\epsilon_0}$ given in the front endpapers,

$$\cot\frac{\theta}{2} = \frac{(5.0\text{ eV})\left(1.602\times 10^{-13}\text{ J/MeV}\right)}{(8.988\times 10^9\text{ N}\cdot\text{m}^2/\text{C}^2)(79)(1.602\times 10^{-19}\text{ C})^2}(2.6\times 10^{-13}\text{ m}) = 11.43,$$

keeping extra significant figures. The scattering angle is then

$$\theta = 2\operatorname{arccot}(11.43) = 2\arctan\left(\frac{1}{11.43}\right) = 10°.$$

4-43: The fraction scattered by less than 1° is $1 - f$, with f given in Equation (4.31);

$$f = \pi n t\left(\frac{Ze^2}{4\pi\epsilon_0\text{ KE}}\right)^2\cot^2\frac{\theta}{2} = \pi n t\left(\frac{1}{4\pi\epsilon_0}\right)^2\left(\frac{Ze^2}{\text{KE}}\right)^2\cot^2\frac{\theta}{2}$$

$$= \pi\left(5.90\times 10^{28}\text{ m}^{-3}\right)\left(3.0\times 10^{-7}\text{ m}\right)\left(8.988\times 10^9\text{ N}\cdot\text{m}^2/\text{C}^2\right)^2$$

$$\times\left(\frac{(79)\left(1.602\times 10^{-19}\text{ C}\right)^2}{(7.7\text{ MeV})\left(1.602\times 10^{-13}\text{ J/MeV}\right)}\right)^2\cot^2(0.5°) = 0.16,$$

where n, the number of gold atoms per unit volume, is from Example 4.8. The fraction scattered by less than 1° is $1 - f = 0.84$.

4-45: Regarding f as a function of θ in Equation (4.31), the number of particles scattered between 60° and 90° is $f(60°) - f(90°)$, and the number scattered through angles greater than 90° is just $f(90°)$, and

$$\frac{f(60°) - f(90°)}{f(90°)} = \frac{\cot^2(30°) - \cot^2(45°)}{\cot^2(45°)} = \frac{3-1}{1} = 2,$$

so twice as many particles are scattered between 60° and 90° than are scattered through angles greater than 90°.

4-47: If gravity acted on photons as if they were massive objects with mass $m = E_\nu/c^2$, the magnitude of the force F in Equation (4.28) would be

$$F = \frac{G M_{\text{sun}} m}{r^2};$$

the factors of r^2 would cancel, as they do for the Coulomb force, and the result is

$$2mc^2 b \sin\frac{\theta}{2} = 2 G M_{\text{sun}} m \cos\frac{\theta}{2} \quad \text{and} \quad \cot\frac{\theta}{2} = \frac{c^2 b}{G M_{\text{sun}}},$$

a result that is independent of the photon's energy. Using $b = R_{\text{sun}}$,

$$\theta = 2 \arctan\left(\frac{G M_{\text{sun}}}{c^2 R_{\text{sun}}}\right) = 2 \arctan\left(\frac{(6.67 \times 10^{-11} \text{ N} \cdot \text{m}^2/\text{kg}^2)(2.0 \times 10^{30} \text{ kg})}{(2.998 \times 10^8 \text{ m/s})^2 (7.0 \times 10^8 \text{ m})}\right)$$

$$= 2.43 \times 10^{-4\circ} = 0.87'',$$

where $1° = 60' = 60$ minutes $= 3600'' = 3600$ seconds.

Chapter 5 - Quantum Mechanics

5-1: Figure (b) is double-valued, and is not a function at all, and cannot have physical significance. Figure (c) has a discontinuous derivative (a "cusp") in the shown interval. Figure (d) is not finite everywhere in the shown interval. Figure (f) is discontinuous in the shown interval.

5-3: The functions (a) and (b) are both infinite when $\cos x = 0$, at $x = \pm\pi/2$, $\pm 3\pi/2, \ldots, \pm(2n+1)\pi/2$ for any integer n, and neither $\psi = A\sec x$ or $\psi = A\tan x$ could be a solution of Schrödginer's equation for all values of x. The function (c) diverges as $x \to \pm\infty$, and cannot be a solution of Schrödinger's equation for all values of x.

5-5: Both parts involve the integral $\int \cos^4 x\, dx$, evaluated between different limits for the two parts. Of the many ways to find this integral, including consulting tables and using symbolic-manipulation programs, a direct algebraic reduction gives

$$\cos^4 x = \left(\cos^2 x\right)^2 = \left[\frac{1}{2}(1+\cos 2x)\right]^2$$
$$= \frac{1}{4}\left[1 + 2\cos 2x + \cos^2(2x)\right]$$
$$= \frac{1}{4}\left[1 + 2\cos 2x + \frac{1}{2}(1+\cos 4x)\right]$$
$$= \frac{3}{8} + \frac{1}{2}\cos 2x + \frac{1}{8}\cos 4x,$$

where the identity $\cos^2\theta = \frac{1}{2}(1+\cos 2\theta)$ has been used twice.

(a) The needed normalization condition is

$$\int_{-\pi/2}^{\pi/2} \psi^*\psi\, dx = A^2 \int_{-\pi/2}^{\pi/2} \cos^4 x\, dx$$
$$= A^2 \left[\frac{3}{8}\int_{-\pi/2}^{\pi/2} dx + \frac{1}{2}\int_{-\pi/2}^{\pi/2}\cos 2x\, dx + \frac{1}{8}\int_{-\pi/2}^{\pi/2}\cos 4x\, dx\right] = 1.$$

The integrals

$$\int_{-\pi/2}^{\pi/2}\cos 2x\, dx = \left.\frac{1}{2}\sin 2x\right|_{-\pi/2}^{\pi/2} \quad \text{and} \quad \int_{-\pi/2}^{\pi/2}\cos 4x\, dx = \left.\frac{1}{4}\sin 4x\right|_{-\pi/2}^{\pi/2}$$

are seen to vanish, and the normalization condition reduces to

$$1 = A^2 \left(\frac{3}{8}\right)\pi, \quad \text{or} \quad A = \sqrt{\frac{8}{3\pi}}.$$

(b) Evaluating the same integral between the different limits,

$$\int_0^{\pi/4} \cos^4 x\, dx = \left[\frac{3}{8}x + \frac{1}{4}\sin 2x + \frac{1}{32}\sin 4x\right]_0^{\pi/4} = \frac{3\pi}{32} + \frac{1}{4}.$$

The probability of the particle being found between $x = 0$ and $x = \pi/4$ is the product of this integral and A^2, or

$$A^2\left(\frac{3\pi}{32} + \frac{1}{4}\right) = \frac{8}{3\pi}\left(\frac{3\pi}{32} + \frac{1}{4}\right) = 0.462.$$

5-7: The given wave function satisfies the continuity condition, and is differentiable to all orders with respect to both t and x, but is not normalizable; specifically, $\Psi^*\Psi = A^*A$ is constant in both space and time, and if the particle is to move freely, there can be no limit to its range, and so the integral of $\Psi^*\Psi$ over an infinite region cannot be finite if $A \neq 0$.

A linear superposition of such waves could give a normalizable wave function, corresponding to a real particle. Such a superposition would necessarily have a non-zero Δp, and hence a finite Δx; at the expense of normalizing the wave function, the wave function is composed of different momentum states, and is localized.

5-9: It's crucial to realize that the expectation value $\langle px \rangle$ is found from the combined operator $\hat{p}\hat{x}$, which, when operating on the wave function $\Psi(x, t)$, corresponds to "multiply by x, differentiate with respect to x and multiply by $\hbar/i$," whereas the operator $\hat{x}\hat{p}$ corresponds to "differentiate with respect to x, multiply by $\hbar/i$ and multiply by x." Using these operators,

$$(\hat{p}\hat{x})\Psi = \hat{p}(\hat{x}\Psi) = \frac{\hbar}{i}\frac{\partial}{\partial x}(x\Psi) = \frac{\hbar}{i}\left[\Psi + x\frac{\partial}{\partial x}\psi\right],$$

where the product rule for partial differentiation has been used. Also,

$$(\hat{x}\hat{p})\Psi = \hat{x}(\hat{p}\Psi) = x\left(\frac{\hbar}{i}\frac{\partial}{\partial x}\Psi\right) = \frac{\hbar}{i}\left[x\frac{\partial}{\partial x}\Psi\right].$$

Thus
$$(\hat{p}\hat{x} - \hat{x}\hat{p})\Psi = \frac{\hbar}{i}\Psi,$$

and
$$\langle px - xp \rangle = \int_{-\infty}^{\infty}\Psi^*\frac{\hbar}{i}\Psi\, dx = \frac{\hbar}{i}\int_{-\infty}^{\infty}\Psi^*\Psi\, dx = \frac{\hbar}{i}$$

for $\Psi(x, t)$ normalized.

5-11: Using $\lambda \nu = v_p$ in Equation (3.5), and using ψ instead of y,

$$\psi = A \cos\left(2\pi\left(t - \frac{x}{v_p}\right)\right) = A \cos\left(2\pi \nu t - 2\pi\frac{x}{\lambda}\right).$$

Differentiating twice with respect to x using the chain rule for partial differentiation (similar to Example 5.1),

$$\frac{\partial \psi}{\partial x} = -A \sin\left(2\pi \nu t - 2\pi\frac{x}{\lambda}\right)\left(-\frac{2\pi}{\lambda}\right) = \frac{2\pi}{\lambda} A \sin\left(2\pi \nu t - 2\pi\frac{x}{\lambda}\right),$$

$$\frac{\partial^2 \psi}{\partial^2 x} = \frac{2\pi}{\lambda} A \cos\left(2\pi \nu t - 2\pi\frac{x}{\lambda}\right)\left(-\frac{2\pi}{\lambda}\right) = -\left(\frac{2\pi}{\lambda}\right)^2 A \cos\left(2\pi \nu t - 2\pi\frac{x}{\lambda}\right)$$

$$= -\left(\frac{2\pi}{\lambda}\right)^2 \psi.$$

The kinetic energy of a nonrelativistic particle is

$$\text{KE} = E - U = \frac{p^2}{2m} = \left(\frac{h}{\lambda}\right)^2 \frac{1}{2m}, \qquad \text{so that}$$

$$\frac{1}{\lambda^2} = \frac{2m}{h^2}(E - U).$$

Substituting the above expression relating $\frac{\partial^2 \psi}{\partial^2 x}$ and $(1/\lambda^2)\,\psi$,

$$\frac{\partial^2 \psi}{\partial^2 x} = -\left(\frac{2\pi}{\lambda}\right)^2 \psi = -\frac{8\pi^2 m}{h^2}(E - U)\psi = -\frac{2m}{\hbar^2}(E - U)\psi,$$

which is Equation (5.32).

5-13: The wave function must vanish at $x = 0$, where $V \to \infty$. As the potential energy increases with x, the particle's kinetic energy must decrease, and so the wavelength increases. The amplitude increases as the wavelength increases because a larger wavelength means a smaller momentum (indicated as well by the lower kinetic energy), and the particle is more likely to be found where the momentum has a lower magnitude. The wave function vanishes again where the potential $V \to \infty$; this condition would determine the allowed energies.

5-15: The necessary integrals are of the form

$$\int_{-\infty}^{\infty} \psi_n \psi_m \, dx = \frac{2}{L} \int_0^L \sin\frac{n\pi x}{L} \sin\frac{m\pi x}{L} \, dx$$

for integers n, m, with $n \neq m$ and $n \neq -m$. (A more general orthogonality relation would involve the integral of $\psi_n^* \psi_m$, but as the eigenfunctions in this problem are real, the distinction need not be made.)

Such integrals are tabulated, or may be found from symbolic-manipulation programs. As an example, the Maple commands that show this result are:

```
>assume(n, integer); additionally(m,integer);
>int(sin(n*Pi*x/L)*sin(m*Pi*x/L),x=0..L);
>int(sin(n*Pi*x/L)*sin(n*Pi*x/L),x=0..L);
```

To do the integrals directly, a convenient identity to use is

$$\sin\alpha \sin\beta = \frac{1}{2}\left[\cos(\alpha - \beta) - \cos(\alpha + \beta)\right],$$

as may be verified by expanding the cosines of the sum and difference of α and β. To show orthogongality, the stipulation $n \neq m$ means that $\alpha \neq \beta$ and $\alpha \neq -\beta$, and the integrals are of the form

$$\int_{-\infty}^{\infty} \psi_n \psi_m \, dx = \frac{1}{L} \int_0^L \left[\cos\frac{(n-m)\pi x}{L} - \cos\frac{(n+m)\pi x}{L}\right] dx$$

$$= \frac{1}{L}\left[\frac{L}{(n-m)\pi}\sin\frac{(n-m)\pi x}{L} - \frac{L}{(n+m)\pi}\sin\frac{(n+m)\pi x}{L}\right]_0^L$$

$$= 0,$$

where $\sin(n-m)\pi = \sin(n-m)\pi = \sin 0 = 0$ has been used.

5-17: Using Equation (5.46), the expectation value $\langle x^2 \rangle$ is

$$\langle x^2 \rangle_n = \frac{2}{L} \int_0^L x^2 \sin^2\left(\frac{n\pi x}{L}\right) dx.$$

Those with access to sybolic-manipulation programs will be able to find the needed definite integral almost immediately. For instance, a possible Maple command is:

```
>assume(n, integer):
 simplify((2/L)*int(x^2*sin(n*Pi*x/L)*sin(n*Pi*x/L),x=0..L));
```

See the end of this chapter for an alternate analytic technique for evaluating this inegral using *Leibniz's Rule*. From either a table or repeated integration by parts, the indefinite integral is

$$\int x^2 \sin^2 \frac{n\pi x}{L} \, dx = \left(\frac{L}{n\pi}\right)^3 \int u^3 \sin u \, du$$

$$= \left(\frac{L}{n\pi}\right)^3 \left[\frac{u^3}{6} - \frac{u^2}{4}\sin 2u - \frac{u}{4}\cos 2u + \frac{1}{8}\sin 2u\right],$$

where the substitution $u = \left(\frac{n\pi}{L}\right)x$ has been made.

This form makes evaluation of the definite integral a bit simpler; when $x = 0$ $u = 0$, and when $x = L$ $u = n\pi$. Each of the terms in the integral vanish at $u = 0$, and the terms with $\sin 2u$ vanish at $u = n\pi$, $\cos 2u = \cos 2n\pi = 1$, and so the result is

$$\langle x^2 \rangle_n = \frac{2}{L}\left(\frac{L}{n\pi}\right)^3 \left[\frac{(n\pi)^3}{6} - \frac{n\pi}{4}\right] = L^2\left[\frac{1}{3} - \frac{1}{2n^2\pi^2}\right].$$

As a check, note that

$$\lim_{n \to \infty} \langle x^2 \rangle_n = \frac{L^2}{3},$$

which is the expectation value of $\langle x^2 \rangle$ in the classical limit, for which the probability distribution is independent of position in the box.

5-19: This is a special case of the probability that such a particle is between x_1 and x_2, as found in Example 5.4. With $x_1 = 0$ and $x_2 = L$,

$$P_{0L} = \left[\frac{x}{L} - \frac{1}{2n\pi}\sin\frac{2n\pi x}{L}\right]_0^L = \frac{1}{n}.$$

5-21: The normalization constant, assuming A to be real, is given by

$$\int \psi^*\psi \, dV = 1 = \int \psi^*\psi \, dx \, dy \, dz$$

$$= A^2 \left(\int_0^L \sin^2 \frac{n_x \pi x}{L} \, dx\right)\left(\int_0^L \sin^2 \frac{n_y \pi y}{L} \, dy\right)\left(\int_0^L \sin^2 \frac{n_z \pi z}{L} \, dz\right).$$

Each integral above is equal to $\frac{L}{2}$ (from calculations identical to Equation (5.43)). The result is

$$A^2 \left(\frac{L}{2}\right)^3 = 1 \quad \text{or} \quad A = \left(\frac{2}{L}\right)^{3/2}.$$

5-23: (a) For the wave function of Problem 5-21, Equation (5.33) must be used to find the energy. Before substitution into Equation (5.33), it is convenient and useful to note that for this wave function

$$\frac{\partial^2 \psi}{\partial^2 x} = -\frac{n_x^2 \pi^2}{L^2} \psi, \quad \frac{\partial^2 \psi}{\partial^2 y} = -\frac{n_y^2 \pi^2}{L^2} \psi, \quad \frac{\partial^2 \psi}{\partial^2 x} = -\frac{n_z^2 \pi^2}{L^2} \psi.$$

Then, substitution into Equation (5.33) gives

$$-\frac{\pi^2}{L^2} \left(n_x^2 + n_y^2 + n_z^2\right) \psi + \frac{2m}{\hbar^2} E\psi = 0,$$

and so the energies are

$$E_{n_x, n_y, n_z} = \frac{\pi^2 \hbar^2}{2mL^2} \left(n_x^2 + n_y^2 + n_z^2\right).$$

(b) The lowest energy occurs when $n_x = n_y = n_z = 1$ (see Problem 5-22). None of the integers n_x, n_y or n_z can be zero, as that would mean $\psi = 0$ identically. The minimum energy is then

$$E_{\min} = \frac{3\pi^2 \hbar^2}{2mL^2},$$

which is three times the ground-state energy of a particle in a one-dimensional box of length L (Equation (5.40) with $n = 1$).

5-25: Solving Equation (5.60) for k_2,

$$k_2 = \frac{1}{2L} \ln \frac{1}{T} = \frac{1}{2(0.200 \times 10^{-9} \text{ m})} \ln(100) = 1.1513 \times 10^{10} \text{ m}^{-1},$$

keeping extra significant figures. Equation (5.86), from the Appendix, may be solved for the energy E, but a more direct expression is

$$E = U - \text{KE} = U - \frac{p^2}{2m} = U - \frac{(\hbar k_2)^2}{2m}$$

$$= 6.00 \text{ eV} - \frac{\left((1.055 \times 10^{-34} \text{ J·s})(1.1513 \times 10^{10} \text{ m}^{-1})\right)^2}{2(9.1095 \times 10^{-31} \text{ kg})(1.602 \times 10^{-19} \text{ J/eV})}$$

$$= 0.949 \text{ eV}.$$

As the potential is given to the nearest 0.01 eV, the electron energy would be known to this precision, or 0.95 eV.

Quantum Mechanics

Special Integrals for Harmonic Oscillators

Many problems in this section involve improper definite integrals of the form

$$\int_{-\infty}^{\infty} y^{2n} e^{-\beta y^2} \, dy$$

for n a nonnegative integer and $\beta > 0$. Integrals of this form are well-known and tabulated, and may be found by use of symbolic-manipulation programs. The general result is used, but not given explicitly, in Equation (5.72). An outline of several methods for finding such integrals is given at the end of the solutions to Chapter 5 in this manual.

By exhibiting Equation (5.72) as a set of normalized wave functions, integrals of the above form may be found. That is, by using $H_n(y)$ as given in Table 5.2,

$$\psi_0^* \psi_0 = \sqrt{\frac{2 m \nu}{\hbar}} e^{-y^2}$$

$$\psi_1^* \psi_1 = \sqrt{\frac{2 m \nu}{\hbar}} \frac{1}{2} (2 y)^2 e^{-y^2}$$

$$\psi_2^* \psi_2 = \sqrt{\frac{2 m \nu}{\hbar}} \frac{1}{8} (4 y^2 - 2)^2 e^{-y^2}.$$

From these expressions it can be seen, for instance, that

$$1 = \int_{-\infty}^{\infty} \psi_0^* \psi_0 \, dx = \sqrt{\frac{2 m \nu}{\hbar}} \sqrt{\frac{\hbar}{2\pi m \nu}} \int_{-\infty}^{\infty} e^{-y^2} \, dy,$$

so that

$$\int_{-\infty}^{\infty} e^{-y^2} \, dy = \sqrt{\pi},$$

where Equation (5.67) has been used to relate x and y, and hence dx and dy.

Similarly,

$$1 = \int_{-\infty}^{\infty} \psi_1^* \psi_1 \, dx = \sqrt{\frac{2 m \nu}{\hbar}} \sqrt{\frac{\hbar}{2\pi m \nu}} \int_{-\infty}^{\infty} y^2 e^{-y^2} \, dy,$$

so that

$$\int_{-\infty}^{\infty} y^2 e^{-y^2} \, dy = \frac{\sqrt{\pi}}{2}.$$

To continue the process with ψ_2,

$$1 = \frac{1}{\sqrt{\pi}} \frac{1}{8} \left[16 \int_{-\infty}^{\infty} y^4 e^{-y^2} \, dy - 16 \int_{-\infty}^{\infty} y^2 e^{-y^2} \, dy + 4 \int_{-\infty}^{\infty} e^{-y^2} \, dy \right]$$

$$= 2 \int_{-\infty}^{\infty} y^4 e^{-y^2} \, dy - 1 + \frac{1}{2},$$

so that

$$\int_{-\infty}^{\infty} y^4 e^{-y^2} \, dy = \frac{3}{4} \sqrt{\pi}$$

5-27: If a particle in a harmonic-oscillator potential had zero energy, the particle would have to be at rest at the position of the potential minimum. The uncertainty principle dictates that such a particle would have an infinite uncertainty in momentum, and hence an infinite uncertainty in energy. This contradiction implies that the zero-point energy of a harmonic oscillator cannot be zero.

5-29: When the classical amplitude of motion is A, the energy of the oscillator is
$$\frac{1}{2} k A^2 = \frac{1}{2} h \nu, \quad \text{so} \quad A = \sqrt{\frac{h\nu}{k}}.$$
Using this for x in Equation (5.67) gives
$$y = \sqrt{\frac{2\pi m \nu}{\hbar}} \sqrt{\frac{h\nu}{k}} = 2\pi \sqrt{\frac{m\nu^2}{k}} = 1,$$
where Equation (5.64) has been used to relate ν, m and k.

5-31: The expectation values will be of the forms
$$\int_{-\infty}^{\infty} x\, \psi^* \psi\, dx \quad \text{and} \quad \int_{-\infty}^{\infty} x^2\, \psi^* \psi\, dx.$$
It is far more convenient to use the dimensionless variable y as defined in Equation (5.67). The necessary integrals will be proportional to
$$\int_{-\infty}^{\infty} y\, e^{-y^2}\, dy, \quad \int_{-\infty}^{\infty} y^2\, e^{-y^2}\, dy, \quad \int_{-\infty}^{\infty} y^3\, e^{-y^2}\, dy \quad \text{and} \quad \int_{-\infty}^{\infty} y^4\, e^{-y^2}\, dy.$$
The first and third integrals are seen to be zero (see Example 5.7). The other two integrals may be found from tables, from symbolic-manipulation programs, or by any of the methods outlined at the end of this chapter or in **Special Integrals for Harmonic Oscillators**, preceding the solutions for Section 5.8 problems in this manual. The integrals are
$$\int_{-\infty}^{\infty} y^2\, e^{-y^2}\, dy = \frac{1}{2}\sqrt{\pi}, \quad \int_{-\infty}^{\infty} y^4\, e^{-y^2}\, dy = \frac{3}{4}\sqrt{\pi}.$$

An immediate result is that $\langle x \rangle = 0$ for the first two states of any harmonic oscillator, and in fact $\langle x \rangle = 0$ for any state of a harmonic oscillator (if $x = 0$ is the minimum of potential energy). A generalization of the above to any case where the potential energy is a symmetric function of x, which gives rise to wave functions that are either symmetric or antisymmetric, leads to $\langle x \rangle = 0$.

To find $\langle x^2 \rangle$ for the first two states, the necessary integrals are

$$\int_{-\infty}^{\infty} x^2\, \psi_0^* \psi_0\, dx = \left(\frac{2m\nu}{\hbar}\right)^{1/2} \left(\frac{\hbar}{2\pi m\nu}\right)^{3/2} \int_{-\infty}^{\infty} y^2 e^{-y^2}\, dy$$

$$= \frac{\hbar}{2\pi^{3/2} m\nu} \frac{\sqrt{\pi}}{2} = \frac{(1/2) h\nu}{4\pi^2 m \nu^2} = \frac{E_0}{k};$$

$$\int_{-\infty}^{\infty} x^2\, \psi_1^* \psi_1\, dx = \left(\frac{2m\nu}{\hbar}\right)^{1/2} \left(\frac{\hbar}{2\pi m\nu}\right)^{3/2} \int_{-\infty}^{\infty} 2 y^4 e^{-y^2}\, dy$$

$$= \frac{\hbar}{2\pi^{3/2} m\nu} 2 \frac{3\sqrt{\pi}}{2} = \frac{(3/2) h\nu}{4\pi^2 m \nu^2} = \frac{E_1}{k}.$$

In both of the above integrals,

$$dx = \frac{dx}{dy}\, dy = \sqrt{\frac{\hbar}{2\pi m\nu}}\, dy$$

has been used, as well as Table 5.2 and Equation (5.64).

5-33: (a) The zero-point energy would be

$$E_0 = \frac{1}{2} h\nu = \frac{h}{2T} = \frac{4.136 \times 10^{-15}\ \text{eV·s}}{2\,(1.00\ \text{s})} = 2.07 \times 10^{-15}\ \text{eV},$$

which is not detectable.

(b) The total energy is $E = mgH$ (here, H is the maximum pendulum height, given as an upper-case letter to distinguish from Planck's constant), and solving Equation (5.70) for n,

$$n = \frac{E}{h\nu} - \frac{1}{2} = \frac{mgH}{h/T}$$

$$= \frac{(1.00 \times 10^{-3}\ \text{kg})\,(9.80\ \text{m/s}^2)\,(1.00 \times 10^{-3}\ \text{m})\,(1.00\ \text{s})}{(6.626 \times 10^{-34}\ \text{J·s})} - \frac{1}{2} = 1.48 \times 10^{28}.$$

Equivalently, using the result of part (a) in place of $h\nu$,

$$n = \frac{1}{2}\left(\frac{E}{E_0} - 1\right) = \frac{1}{2}\left(\frac{mgH}{E_0} - 1\right)$$

$$= \frac{1}{2}\left(\frac{(1.00 \times 10^{-3}\ \text{kg})\,(9.80\ \text{m/s}^2)\,(1.00 \times 10^{-3}\ \text{m})\,(1.00\ \text{s})}{(2.07 \times 10^{-15}\ \text{eV})\,(1.602 \times 10^{-19}\ \text{J/eV})} - 1\right)$$

$$= 1.48 \times 10^{28}.$$

For **Problems 34, 35 and 36**, it is most convenient to use the wave functions in terms of the dimensionless variable y, as given in Equations (5.67) and (5.72), intead of x. The normalization constants need not be considered in showing that the ψ_n are solutions of Schrödinger's equation as given in Equation (5.69). These exercises reduce to showing that the functions

$$H_n(y)\, e^{-y^2},$$

with $H_n(y)$ as given in Table 5.2, are solutions to Equation (5.69).

A commonly-appearing differentiation is the derivative with respect to y of the product of a polynomial $P(y)$ in y and e^{-y^2};

$$\frac{d}{dy}\left(P(y)\, e^{-y^2}\right) = \left(\frac{d}{dy}P(y) - y\,P(y)\right) e^{-y^2}.$$

5-35: The unnormalized wave function is $\psi_2 = \left(2y^2 - 1\right) e^{-y^2}$, and

$$\frac{d}{dy}\psi_2 = \left(4y - 2y^3 + y\right) e^{-y^2}, \qquad \frac{d^2}{dy^2}\psi_2 = \left(5 - 6y^2 - 5y^2 + 2y^4\right) e^{-y^2}.$$

Combining powers of y,

$$\frac{d^2}{dy^2}\psi_2 - y^2\,\psi_2 = \left(5 - 11y^2 + 2y^4 - 2y^4 + y^2\right) e^{-y^2} = \left(5 - 10y^2\right) e^{-y^2} = -5\psi_2,$$

and so ψ_2 is a solution to Equation (5.69) with $\alpha = 5$.

5-37: (a) In the notation of the Appendix, the wave function in the two regions has the form

$$\psi_\mathrm{I} = A\,e^{i k_1 x} + B\,e^{-i k_1 x}, \qquad \psi_\mathrm{II} = C\,e^{i k' x} + D\,e^{-i k' x},$$

where
$$k_1 = \sqrt{\frac{2mE}{\hbar}}, \qquad k' = \sqrt{\frac{2m(E-U)}{\hbar}}.$$

The terms corresponding to $e^{-i k_1 x}$ and $e^{-i k' x}$ represent particles traveling to the left; this is possible in region I, due to reflection at the step at $x = 0$, but not in region II (the reasoning is the same as that which lead to setting $G = 0$ in Equation (5.82)). Therefore, the $e^{-i k' x}$ term is not physically meaningful, and $D = 0$.

Quantum Mechanics

(b) The boundary conditions at $x = 0$ are then

$$A + B = C, \qquad i k_1 A - i k_1 B = i k' C \quad \text{or} \quad A - B = \frac{k'}{k_1} C.$$

Adding to eliminate B, $2A = \left(1 + \dfrac{k'}{k_1}\right) C$, so

$$\frac{C}{A} = \frac{2 k_1}{k_1 + k'}, \qquad \text{and} \qquad \frac{CC^*}{AA^*} = \frac{4 k_1^2}{(k_1 + k')^2}.$$

(Note that the ratios C/A and C^*/A^* are real in this case.)

(c) The particle speeds are different in the two regions, so Equation (5.83) becomes

$$T = \frac{|\psi_{\text{II}}|^2 v'}{|\psi_{\text{I}}|^2 v_1} = \frac{CC^*}{AA^*} \frac{k'}{k_1} = \frac{4 k_1 k'}{(k_1 + k')^2} = \frac{4 (k_1/k')}{((k_1/k') + 1)^2}.$$

For the given situation, $k_1/k' = v_1/v' = 2.00$, so $T = \dfrac{(4)(2)}{((2)+1)^2} = \dfrac{8}{9}$. The transmitted current is $(T)(1.00 \text{ mA}) = 0.889$ mA, and the reflected current is 0.111 mA.

As a check on the last result, note that the ratio of the reflected current to the incident current is, in the notation of the Appendix,

$$R = \frac{|\psi_{\text{I}+}|^2 v_1}{|\psi_{\text{I}+}|^2 v_1} = \frac{BB^*}{AA^*}.$$

Eliminating C from the equations obtained in part (b) from the continuity condition as $x = 0$,

$$A\left(1 - \frac{k'}{k_1}\right) = B\left(1 + \frac{k'}{k_1}\right), \qquad \text{so}$$

$$R = \left(\frac{(k_1/k') - 1}{(k_1/k') + 1}\right)^2 = \frac{1}{9} = 1 - T,$$

as expected.

A Further Examination of Integrals for the Harmonic Oscillator

Many problems in this chapter involve improper definite integrals of the form

$$\int_{-\infty}^{\infty} y^{2n} e^{-\beta y^2} \, dy$$

for n a nonnegative integer and $\beta > 0$. Integrals of this form are well-known and tabulated, and may be found by use of symbolic-manipulation programs. The general result is used, but not given eyplicitly, in Equation (5.72).

While consulting tables or programs is certainly adequate for most purposes, there is some advantage in seeing how these integrals are obtained.

The starting point is often consideration of the integral

$$\int_{-\infty}^{\infty} e^{-u^2} \, du = \sqrt{\pi},$$

a standard result obtained from the transformation of a double integral from cartesian to polar coordinates; the derivation will not be reproduced here. However, making the substitution $u = \sqrt{\beta}\, y$ leads, after a basic change of variable, to

$$\int_{-\infty}^{\infty} e^{-\beta y^2} \, dy = \sqrt{\frac{\pi}{\beta}}.$$

Integration by parts may be used to obtain a recurrence relation. In the above relation,

$$\int_{-\infty}^{\infty} e^{-u^2} \, du = u\, e^{-u^2} \Big|_{-\infty}^{\infty} - \int_{-\infty}^{\infty} u\, d\!\left(e^{-u^2}\right) = 0 + 2 \int_{-\infty}^{\infty} u^2 e^{-u^2} \, du,$$

or

$$\int_{-\infty}^{\infty} u^2 e^{-u^2} \, du = \frac{1}{2} \int_{-\infty}^{\infty} e^{-u^2} \, du = \frac{\sqrt{\pi}}{2}.$$

Similarly (skipping the explicit inclusion of the boundary terms),

$$\int_{-\infty}^{\infty} u^2 e^{-u^2} \, du = \int_{-\infty}^{\infty} e^{-u^2} \, d\!\left(\frac{u^3}{3}\right) = \frac{1}{3} \int_{-\infty}^{\infty} u^3 \, d\!\left(e^{-u^2}\right) = \frac{2}{3} \int_{-\infty}^{\infty} u^4 e^{-u^2} \, du,$$

so that

$$\int_{-\infty}^{\infty} u^4 e^{-u^2} \, du = \frac{3\sqrt{\pi}}{4}.$$

A pattern soon emerges, and it may be seen by induction that

$$\int_{-\infty}^{\infty} u^{2n} e^{-u^2} \, du = \frac{(2n-1)(2n-3)\cdots(3)(1)}{2^n} \sqrt{\pi}.$$

To see this result using Maple, a possible set of commands are:

```
>assume(n,integer); additionally(n>0):
>f:=u^(2*n)*exp(-u^2);
>int(f,u=-infinity..infinity);
```

However, this result, in terms of a Gamma function, is often not useful. To see the above pattern for small integral values of n, the command (but don't enter linebreaks!)

```
>for m from 1 to 10 do
    int(u^(2*m)*exp(-u^2),u=-infinity..infinity) od;
```

shows the above pattern.

Making the same substitution $u = \sqrt{\beta}\, y$ yields

$$\int_{-\infty}^{\infty} y^{2n} e^{-\beta y^2}\, dy = \frac{(2n-1)(2n-3)\cdots(3)(1)}{2^n} \sqrt{\frac{\pi}{\beta^{2n+1}}}.$$

Those familiar with *Leibniz's Formula* will recognize that

$$\int_{-\infty}^{\infty} y^{2n} e^{-\beta y^2}\, dy = (-1)^n \left(\frac{d^n}{d\beta^n}\right) \int_{-\infty}^{\infty} e^{-\beta y^2}\, dy = (-1)^n \left(\frac{d^n}{d\beta^n}\right) \sqrt{\frac{\pi}{\beta}},$$

the same result obtained by an equivalent and possibly simpler (in terms of calculation) method.

The needed integrals for the first few values of n are:

$n = 1$ $\qquad\displaystyle\int_{-\infty}^{\infty} e^{-\beta y^2}\, dy = \sqrt{\frac{\pi}{\beta}}$

$n = 2$ $\qquad\displaystyle\int_{-\infty}^{\infty} y^2 e^{-\beta y^2}\, dy = \frac{1}{2}\sqrt{\frac{\pi}{\beta^3}}$

$n = 3$ $\qquad\displaystyle\int_{-\infty}^{\infty} y^4 e^{-\beta y^2}\, dy = \frac{3}{4}\sqrt{\frac{\pi}{\beta^5}}$

For physics applications, it can never hurt to check the dimensions; if y has dimensions of length, β must have dimensions of $[\text{length}]^{-2}$, and $\int_{-\infty}^{\infty} y^{2n} e^{-\beta y^2}\, dy$ must have dimensions of $[\text{length}]^{2n+1}$. In the results presented above, the integral has the same dimensions as $\beta^{-(2n+1)/2}$, and the dimensions are seen to be consistent.

A Further Examination of Integrals related to a Particle in a Box

As an alternative to finding the indefinite integral

$$\langle x^2 \rangle_n = \frac{2}{L} \int \sin^2\left(\frac{n\pi x}{L}\right) dx,$$

it is interesting to note that the simpler definite integral $\int_0^L \sin^2(n\pi x/) \, dx$ may be used to find the definite integral $\int_0^L x^2 \sin^2(n\pi x/) \, dx$ by letting $\alpha = n\pi/L$ be a parameter and using *Leibniz's Rule* to differentiate the integral *with respect to α*. That is,

$$\int_0^L \sin^2 \frac{n\pi x}{L} \, dx = \int_0^L \sin^2 \alpha x \, dx = \frac{L}{2} = \frac{n\pi}{2\alpha},$$

or

$$\frac{1}{2} \int_0^L (1 - \cos 2\alpha x) \, dx = \frac{n\pi}{2\alpha}$$

Then, differentiating both sides of this last relation with respect to α twice,

$$-\int_0^L x \sin 2\alpha x \, dx = -\frac{n\pi}{2\alpha^2}, \qquad -2\int_0^L x^2 \cos 2\alpha x \, dx = \frac{n\pi}{\alpha^3}.$$

(In the two expressions above, there are terms corresponding to, respectively,

$$\frac{dL}{d\alpha} L \sin 2\alpha L \qquad \text{and} \qquad \frac{dL}{d\alpha} L^2 \cos 2\alpha L,$$

but thes both vanish.)

Using $\cos 2\alpha x = 1 - 2\sin^2 \alpha x$ gives

$$-2\int_0^L (1 - 2\sin^2 \alpha x) \, dx = -\frac{2}{3} L^3 + 4 \int_0^L x^2 \sin^2 \frac{n\pi x}{L} \, dx = \frac{n\pi}{\alpha^3} = \frac{L^3}{n^2 \pi^2},$$

from which the previous result is obtatined,

$$\frac{2}{L} \int_0^L x^2 \sin^2 \frac{n\pi x}{L} \, dx = L^2 \left[\frac{1}{3} - \frac{1}{2 n^2 \pi^2} \right].$$

Chapter 6 - Quantum Theory of the Hydrogen Atom

6-1: Whether in cartesian (x, y, z) or spherical coordinates, three quantities are needed to describe the variation of the wave function throughout space. The three quantum numbers needed to describe an atomic electron correspond to the variation in the radial direction, the variation in the azimuthal direction (the variation along the circumference of the classical orbit), and the variation with the polar direction (variation along the direction from the classical axis of rotation).

6-3: For the given function,

$$\frac{d}{dr} R_{10} = -\frac{2}{a_0^{5/2}} e^{-r/a_0}, \quad \text{and}$$

$$\frac{1}{r^2} \frac{d}{dr}\left(r^2 \frac{dR_{10}}{dr}\right) = -\frac{2}{a_0^{5/2}} \frac{1}{r^2}\left(2r - \frac{r^2}{a_0}\right) e^{-r/a_0}$$

$$= \left(\frac{1}{a_0^2} - \frac{2}{r\, a_0}\right) R_{10}.$$

This is a solution to Equation (6.14) if $l = 0$ (as indicated by the index of R_{10}),

$$\frac{2}{a_0} = \frac{2\, m\, e^2}{\hbar^2\, 4\pi\, \epsilon_0}, \quad \text{or} \quad a_0 = \frac{4\pi^2\, \epsilon_0\, \hbar^2}{m\, e^2},$$

which is the case, and

$$\frac{2m}{\hbar^2} E = -\frac{1}{a_0^2}, \quad \text{or} \quad E = -\frac{e^2}{8\pi\, \epsilon_0\, a_0} = E_1,$$

again as indicated by the index of R_{10}.

To show normalization,

$$\int_0^\infty |R_{10}|^2\, r^2\, dr = \frac{4}{a_0^3} \int_0^\infty r^2 e^{-2r/a_0}\, dr = \frac{1}{2} \int_0^\infty u^2 e^{-u}\, du,$$

where the substitution $u = 2r/a_0$ has been made. The improper definite integral in u is known to have the value 2 (see the discussion at the end of this chapter), and so the given function is normalized.

6-5: From Equation (6.15) the integral, apart from the normalization constants, is

$$\int_0^{2\pi} \Phi_{m_l}^* \Phi_{m_l'} \, d\phi = \int_0^{2\pi} e^{-i\,m_l\,\phi} e^{i\,m_l'\,\phi} \, d\phi.$$

It is possible to express the integral in terms of real and imaginary parts, but it turns out to be more convenient to do the integral directly in terms of complex exponentials;

$$\int_0^{2\pi} e^{-i\,m_l\,\phi} e^{i\,m_l'\,\phi} \, d\phi = \int_0^{2\pi} e^{i\,(m_l' - m_l)\,\phi} \, d\phi$$

$$= \frac{1}{i\,(m_l' - m_l)} \left[e^{i\,(m_l' - m_l)\,\phi} \right]_0^{2\pi} = 0.$$

The above form for the integral is valid only for $m_l \neq m_l'$, which is given for this case. In evaluating the integral at the limits, the fact that $e^{i\,2\pi\,n} = 1$ for any integer n (in this case $(m_l' - m_l)$) has been used.

6-7: In the Bohr model, for the ground-state orbit of an electron in a hydrogen atom, $\lambda = \dfrac{h}{mv} = 2\pi r$, and so $L = pr = \hbar$. In the quantum theory, zero-angular-momentum states (ψ spherically symmetric) are allowed, and $L = 0$ for a ground-state hydrogen atom.

6-9: From Equation (6.22), L_z must be an integer multiple of $\hbar$; for L to be equal to L_z, the product $l(l+1)$, from Equation (6.21), must be the square of some integer less than or equal to l. But,

$$l^2 \leq l(l+1) < (l+1)^2$$

for any nonnegative l, with equality holding in the first relation only if $l = 0$. Therefore, $l(l+1)$ is the square of an integer only if $l = 0$, in which case $L_z = 0$ and $L = L_z = 0$.

6-11: From Equation (6.22), the possible values for the magnetic quantum number m_l are

$$m_l = 0, \pm 1, \pm 2, \pm 3, \pm 4,$$

a total of nine possible values.

Quantum Theory of the Hydrogen Atom   57

6-13: The fractional difference between L and the largest value of L_z is, for a given l,

$$\frac{L - L_{z,\max}}{L} = \frac{\sqrt{l(l+1)} - l}{\sqrt{l(l+1)}} = 1 - \frac{l}{\sqrt{l(l+1)}} = 1 - \sqrt{\frac{l}{l+1}}.$$

For a p state, $l = 1$ and $1 - \sqrt{\frac{1}{2}} = 0.29 = 29\%$.

For a d state, $l = 2$ and $1 - \sqrt{\frac{2}{3}} = 0.18 = 18\%$.

For an f state, $l = 3$ and $1 - \sqrt{\frac{3}{4}} = 0.13 = 13\%$.

6-15: Using $R_{10}(r)$ from Table 6.1 in Equation (6.25),

$$P(r) = \frac{4\,r^2}{a_0^3}\,e^{-2r/a_0}.$$

The most probable value of r is that for which $P(r)$ is a maximum. Differentiating the above expression for $P(r)$ with respect to r and setting the derivative equal to zero,

$$\frac{d}{dr}P(r) = \frac{4}{a_0^3}\left(2r - \frac{2r^2}{a_0}\right)e^{-2r/a_0} = 0, \quad \text{or}$$

$$r = \frac{r^2}{a_0} \quad \text{and} \quad r = 0,\ a_0$$

for an extreme. At $r = 0$, $P(r) = 0$, and because $P(r)$ is never negative, this must be a minimum. $\dfrac{dP}{dr} \to 0$ as $r \to \infty$, and this also corresponds to a minimum. The only maximum of $P(r)$ is at $r = a_0$, which is the radius of the first Bohr orbit.

6-17: Using $R_{21}(r)$ from Table 6.1 in Equation (6.25), and ignoring the leading constants (which would not affect the position of extremes),

$$P(r) = r^6\,e^{-2r/3a_0}.$$

The most probable value of r is that for which $P(r)$ is a maximum. Differentiating the above expression for $P(r)$ with respect to r and setting the derivative equal to zero,

$$\frac{d}{dr}P(r) = \left(6r^5 - \frac{2r^6}{3\,a_0}\right)e^{-2r/3a_0} = 0, \quad \text{or}$$

$$6r^5 = \frac{2r^6}{3\,a_0} \quad \text{and} \quad r = 0,\ 9a_0$$

for an extreme. At $r = 0$, $P(r) = 0$, and because $P(r)$ is never negative, this must be a minimum. $\dfrac{dP}{dr} \to 0$ as $r \to \infty$, and this also corresponds to a minimum. The only maximum of $P(r)$ is at $r = 9\,a_0$, which is the radius of the third Bohr orbit.

6-19: For the ground state, $n = 1$, the wave function is independent of angle, as seen from the functions $\Phi(\phi)$ and $\Theta(\theta)$ in Table 6.1, where for $n = 1$, $l = m_l = 0$ (see Problem 6-14). The ratio of the probabilities is then the ratio of the product $r^2 (R_{10}(r))^2$ evaluated at the different distances. Specially,

$$\frac{P(a_0)\, dr}{P(a_0/2)\, dr} = \frac{(a_0)^2\, e^{-2a_0/a_0}}{(a_0/2)^2\, e^{-2(a_0/2)/a_0}} = \frac{e^{-2}}{(1/4)\, e^{-1}} = \frac{4}{e} = 1.47.$$

Similarly,

$$\frac{P(a_0)\, dr}{P(2\, a_0)\, dr} = \frac{(a_0)^2\, e^{-2a_0/a_0}}{(2\, a_0)^2\, e^{-2(2a_0)/a_0}} = \frac{e^{-2}}{4\, e^{-4}} = \frac{e^2}{4} = 1.85.$$

6-21: (a) Using $R_{10}(r)$ for the $1s$ radial function from Table 6.1,

$$\int_{a_0}^{\infty} |R(r)|^2\, r^2\, dr = \frac{4}{a_0^3} \int_{a_0}^{\infty} r^2\, e^{-2r/a_0}\, dr = \frac{1}{2} \int_{2}^{\infty} u^2\, e^u\, du,$$

where the substitution $u = 2r/a_0$ has been made.

There are numerous ways to find the definite integral, including consulting tabulated integrals or using symbolic-manipulation program; for example, the Maple command

>int((1/2)*u^2*exp(-u),u=2..infinity);

gives the result immediately. Using the method outlined at the end of this chapter to find the improper definite integral leads to

$$\frac{1}{2} \int_{2}^{\infty} u^2\, e^u\, du = \frac{1}{2} \left[-e^{-u} (u^2 + 2u + 2) \right]_{2}^{\infty} = \frac{1}{2} \left[e^{-2}\, 10 \right] = \frac{5}{e^2} = 0.68 = 68\%.$$

(b) Repeating the above calculation with $2\, a_0$ as the lower limit of the integral,

$$\frac{1}{2} \int_{4}^{\infty} u^2\, e^u\, du = \frac{1}{2} \left[-e^{-u} (u^2 + 2u + 2) \right]_{4}^{\infty} = \frac{1}{2} \left[e^{-4}\, 26 \right] = \frac{13}{e^4} = 0.24 = 24\%.$$

6-23: For $l = 0$, only $m_l = 0$ is allowed, $\Phi(\phi)$ and $\Theta(\theta)$ are both constants (from Table 6.1)), and the theorem is verified.

For $l = 1$, the sum is

$$\frac{1}{2\pi} \frac{3}{4} \sin^2 \theta + \frac{1}{2\pi} \frac{3}{2} \cos^2 \theta + \frac{1}{2\pi} \frac{3}{4} \sin^2 \theta = \frac{3}{4\pi}.$$

Quantum Theory of the Hydrogen Atom   **59**

In the above, $\Phi^*\Phi = \dfrac{1}{2\pi}$, which holds for any l and m_l, has been used. Note that one term appears twice, one for $m_l = -1$ and once for $m_l = 1$.

For $l = 2$, combining the identical terms for $m_l = \pm 2$ and $m_l = \pm 1$, and again using $\Phi^*\Phi = \dfrac{1}{2\pi}$, the sum is

$$2\,\frac{1}{2\pi}\,\frac{15}{16}\sin^4\theta + 2\,\frac{1}{2\pi}\,\frac{15}{4}\sin^2\theta\cos^2\theta + \frac{1}{2\pi}\,\frac{10}{16}\left(3\cos^2\theta - 1\right)^2.$$

The above may be simplified by extracting the commons constant factors, to

$$\frac{5}{16\pi}\left[(3\cos^2\theta - 1)^2 + 12\sin^2\theta\cos^2\theta + 3\sin^4\theta\right].$$

Of the many ways of showing the term in brackets is indeed a constant, the one presented here, using a bit of hindsight, seems to be one of the more direct methods. Using the identity $\sin^2\theta = 1 - \cos^2\theta$ to elminate $\sin\theta$,

$$(3\cos^2\theta - 1)^2 + 12\sin^2\theta\cos^2\theta + 3\sin^4\theta$$
$$= \left(9\cos^4\theta - 6\cos^2\theta + 1\right) + 12\left(1 - \cos^2\theta\right)\cos^2\theta + 3\left(1 - 2\cos^2\theta + \cos^4\theta\right)$$
$$= 1,$$

and the theorem is verified.

6-25: In the integral of Equation (6.35), the radial integral will never vanish, and only the angular functions $\Phi(\phi)$ and $\Theta(\theta)$ need to be considered. The $\Delta l = 0$ transition is seen to be forbidden, in that the product

$$\left(\Phi_0(\phi)\,\Theta_{00}(\theta)\right)^*\left(\Phi_0(\phi)\,\Theta_{00}(\theta)\right) = \frac{1}{4\pi}$$

is spherically symmetric, and any integral of the form of Equation (6.35) must vanish, as the argument $u = x$, y or z will assume positive and negative values with equal probability amplitudes.

If $l = 1$ in the initial state, the integral in Equation (6.35) will be seen to to vanish if u is chosen appropriately. If $m_l = 0$ initially, and $u = z = r\cos\theta$ is used, the integral (apart from constants) is

$$\int_0^\pi \cos^2\theta\,\sin\theta\,d\theta = \frac{2}{3} \neq 0.$$

If $m_l = \pm 1$ initially, and $u = x = r\sin\theta\cos\phi$ is used, the θ-integral is of the form

$$\int_0^\pi \sin^2\theta\,d\theta = \frac{\pi}{2} \neq 0$$

and the ϕ-integral is of the form

$$\int_0^{2\pi} e^{\pm i\phi}\cos\phi\,d\phi = \int_0^{2\pi} \cos^2\phi\,d\phi = \pi \neq 0,$$

and the transition is allowed.

6-27: The relevant integrals are of the form

$$\int_0^L x \sin\frac{n\pi x}{L} \sin\frac{m\pi x}{L}\, dx.$$

The integrals may be found in a number of ways, including consulting tables or using symbolic-manipulation programs (see, for instance, the solution to Problem 5-15 for sample Maple commands that are easily adpated to this problem).

One way to find a general form for the integral is to use the identity

$$\sin\alpha\,\sin\beta = \frac{1}{2}\left[\cos(\alpha-\beta) - \cos(\alpha+\beta)\right]$$

and the indefinite integral (found from integration by parts)

$$\int x\cos kx\, dx = \frac{x\sin kx}{k} - \frac{1}{k}\int \sin kx\, dx = \frac{x\sin kx}{k} + \frac{\cos kx}{k^2}$$

to find the above definite integral as

$$\frac{1}{2}\left[\frac{Lx}{(n-m)\pi}\sin\frac{(n-m)\pi x}{L} + \frac{L^2}{(n-m)^2\pi^2}\cos\frac{(n-m)\pi x}{L} \right.$$
$$\left. -\frac{Lx}{(n+m)\pi}\sin\frac{(n+m)\pi x}{L} + \frac{L^2}{(n+m)^2\pi^2}\cos\frac{(n+m)\pi x}{L}\right]_0^L,$$

where $n^2 \neq m^2$ is assumed. The terms involving sines vanish, with the result of

$$\frac{L^2}{2\pi^2}\left[\frac{\cos(n-m)\pi - 1}{(n-m)^2} - \frac{\cos(n+m)\pi - 1}{(n+m)^2}\right].$$

If n and m are both odd or both even, $n+m$ and $n-m$ are even, the arguments of the cosine terms in the above expression are even-integral multiples of π, and the integral vanishes. Thus, the $n=3 \to n=1$ transition is forbidden, while the $n=3 \to n=2$ and $n=2 \to n=1$ transitions are allowed.

To make use of symmetry arguments, consider that

$$\int_0^L \left(x - \frac{L}{2}\right)\sin\frac{n\pi x}{L}\sin\frac{m\pi x}{L}\, dx = \int_0^L x\sin\frac{n\pi x}{L}\sin\frac{m\pi x}{L}\, dx$$

for $n \neq m$, because the integral of L times the product of the wave functions is zero; the wave functions were shown to be orthogonal in Chapter 5 (again, see Problem 5-15). Letting $u = \frac{L}{2} - x$,

$$\sin\frac{n\pi x}{L} = \sin\frac{n\pi((L/2)-u)}{L} = \sin\left(\frac{n\pi}{2} - \frac{n\pi u}{L}\right).$$

This expression will be $\pm\cos\left(\frac{n\pi u}{L}\right)$ for n odd and $\pm\sin\left(\frac{n\pi u}{L}\right)$ for n even. The integrand is then an odd function of u when n and m are both even or both odd, and hence the integral is zero. If one of n or m is even and the other odd, the integrand is an even function of u and the integral is nonzero.

Quantum Theory of the Hydrogen Atom

6-29: From Equation (6.39), the magnitude of the magnetic moment of an electron in a Bohr orbit is porportional to the magnitude of the angular momentum, and hence proportional to n. The orbital radius is proportional to n^2 (see Equation (4.13) or Problem 4-28), and so the magnetic moment is proportional to $\sqrt{r_n}$.

6-31: See Example 6.4; solving for B,

$$B = \frac{\Delta\lambda}{\lambda^2}\frac{4\pi m c}{e}$$

$$= \frac{(0.010 \times 10^{-9}\text{ m})}{(400 \times 10^{-0}\text{ m})^2}\frac{4\pi (9.1095 \times 10^{-31}\text{ kg})(2.998 \times 10^8\text{ m/s})}{(1.602 \times 10^{-19}\text{ C})} = 1.34\text{ T}.$$

Evaluation of Integrals for Hydrogen Wave Functions

Many of the problems in this chapter involve integrals of the form

$$\int u^n e^{-u}\, du \quad\text{or}\quad \int_0^\infty u^n e^{-u}\, du$$

for n a nonnegative integer.

Such integrals are well-known and may be found from tables or by use of symbolic-manipulation programs. Standard methods of finding these integrals involve repeated integration by parts. Specifically, for $n = 0$, the idefinite integral is readily found,

$$\int e^{-u}\, du = -e^{-u}$$

and the improper definite integral is seen to be

$$\int_0^\infty e^{-u}\, du = 1.$$

Integrating the above indefinite integral by parts,

$$\int e^{-u}\,du = u\,e^{-u} - \int u\,d\left(-e^{-u}\right)$$
$$= u\,e^{-u} + \int u\,e^{-u}\,du,$$

so that
$$\int u\,e^{-u}\,du = -e^{-u}(u+1).$$

The improper definite integral is seen to be

$$\int_0^\infty u\,e^{-u}\,du = 1.$$

The process may be repeated, or the process of integration by parts may be generalized as

$$\int u^n\,e^{-u}\,du = u^{n+1}\,e^{-u}\,du + n\int u^{n+1}\,e^{-u}\,du.$$

The first few such integrals are:

$n = 0$ $\qquad \int e^{-u}\,du = -e^{-u}$

$n = 1$ $\qquad \int u\,e^{-u}\,du = -e^{-u}(u+1)$

$n = 2$ $\qquad \int u^2\,e^{-u}\,du = -e^{-u}(u^2+2u+2)$

$n = 3$ $\qquad \int u^3\,e^{-u}\,du = -e^{-u}(u^3+3u^2+6u+6)$

$n = 4$ $\qquad \int u^4\,e^{-u}\,du = -e^{-u}(u^4+4u^3+12u^2+24u+24)$

Further integrals may be found by, for instance, variations on the Maple command

>for m from 0 to 10 do int(u^m*exp(-u),u) od;

For the definite improper integrals, a pattern, easily verified by induction is found, the standard result

$$\int_0^\infty u^n\,e^{-u}\,du = n!.$$

The Maple command that would demonstrate this pattern is

>for n from 0 to 10 do int(u^n*exp(-u),u=0..infinity) od;

Chapter 7 - Many-Electron Atoms

7-1: (a) Using Equations (7.4) and (6.41), the energy difference is

$$\Delta E = 2\mu_{sz} B = 2\mu_B B = 2\left(5.788 \times 10^{-5} \text{ eV/T}\right)(1.20 \text{ T}) = 1.39 \times 10^{-4} \text{ eV}.$$

(b) The wavelength of the radiation that corrsponds to this energy is

$$\lambda = \frac{hc}{\Delta E} = \frac{1.240 \times 10^{-6} \text{ eV·m}}{1.389 \times 10^{-4} \text{ eV}} = 8.93 \text{ mm}.$$

Note that a more precise value of ΔE was needed in the intermediate calculation to avoid roundoff error.

7-3: For an electron, $s = \left(\sqrt{3}/2\right)\hbar$, $s_z = \pm(1/2)\hbar$, and so the possible angles are given by

$$\arccos\left(\frac{\pm(1/2)\hbar}{\left(\sqrt{3}/2\right)\hbar}\right) = \arccos\left(\pm\frac{1}{\sqrt{3}}\right) = 54.7°, 125.3°.$$

7-5: $^4_2$He atoms contain even numbers of spin-$\frac{1}{2}$ particles, which pair off to give zero or integral spins for the atoms. Such atoms do not obey the exclusion principle. $^3_2$He atoms contain odd numbers of spin-$\frac{1}{2}$ particles, and so have net spins of $\frac{1}{2}$, $\frac{3}{2}$ or $\frac{5}{2}$, and they obey the exclusion principle.

7-7: An alkali metal atom has one electron outside closed inner shells: A halogen atom lacks one electron of having a closed outer shell: An inert gas atom has a closed outer shell.

7-9: For and f subshell, with $l = 3$, the possbile values of m_l are ± 2, ± 2, ± 1 or 0, for a total of $2l+1 = 7$ values of m_l. Each state can have two electrons of opposite spins, for a total of 14 electrons.

7-11: The number of elements would be the total number of electrons in all of the shells. Repeated use of Equation (7.14) gives

$$2n^2 + 2(n-1)^2 + \cdots + 2(1)^2 = 2(36 + 25 + 16 + 9 + 4 + 1) = 182.$$

In general, using the expression for the sum of the squares of the first n integers, the number of elements would be

$$2\left(\frac{1}{6}n(2n+1)(n+1)\right) = \frac{1}{3}(n(2n+1)(n+1)),$$

which gives a total of 182 elements when $n = 6$.

7-13: All of the atoms are hydrogenlike, in that there is a completely filled subshell that screens the nuclear charge and causes the atom to "appear" to be a single charge. The outermost electron in each of these atoms is further from the nucleus for higher atomic number, and hence has a successively lower binding energy.

7-15: (a) See Table 7.4. The $3d$ subshell is empty, and so the effective nuclear charge is roughly $+2e$, and the outer electron is relatively easy to detach.

(b) Again, see Table 7.4. The completely filled K and L shells shield $+10e$ of the nuclear charge of $= 16e$; the filled $3s^2$ subshell will partially shield the nuclear charge, but not to the same extent as the filled shells, so $+6e$ is a rough estimate for the effective nuclear charge. This outer electron is then relatively hard to detach.

7-17: Cl^- ions have closed shells, whereas a Cl atom is one electron short of having a closed shell and the relatively poorly shielded nuclear charge tends to attract an electron from another atom to fill the shell.

Na^+ ions have closed shells, whereas an Na atom has a single outer electron that can be detached relatively easily in a chemical reaction with another atom.

7-19: The Li atom ($Z = 3$) is larger because the effective nuclear charge acting on its outer electron is less than that acting on the outer electrons of the F atom ($Z = 9$). The Na atom ($Z = 11$) is larger because it has an additional electron shell (see Table 7.4). The Cl atom ($Z = 17$) atom is larger beacuse has an additional electron shell. The Na atom is larger than the Si atom ($Z = 14$) for the same reason as given for the Li atom.

7-21: The only way to produce a normal Zeeman effect is to have no net electron spin; because the electron spin is $\pm\frac{1}{2}$, the total number of electrons must be even. If the total number of electrons were odd, the net spin would be nonzero, and the anamolous Zeeman effect would be observable.

7-23: See Example 7.6. Expressing the difference in energy levels as

$$\Delta E = 2\mu_B B = hc\left(\frac{1}{\lambda_1} - \frac{1}{\lambda_2}\right); \quad \text{solving for } B,$$

$$B = \frac{hc}{2\mu_B}\left(\frac{1}{\lambda_1} - \frac{1}{\lambda_2}\right)$$

$$= \frac{(1.240 \times 10^{-6} \text{ eV·m})}{2(5.788 \times 10^{-5} \text{ eV/T})}\left(\frac{1}{589.0 \times 10^{-9} \text{ m}} - \frac{1}{589.6 \times 10^{-9} \text{ m}}\right) = 18.5 \text{ T.}$$

Many-Electron Atoms 65

7-25: The possible values of l are $j + \frac{1}{2} = 3$ and $j - \frac{1}{2} = 2$.

7-27: For the ground state to be a singlet state with no net angular momentum, all of the subshells must be filled.

7-29: For this doublet state, **L** $= 0$, **S** $=$ **J** $= \frac{1}{2}$. There are no other allowed states. This state has the lowest possible values of **L** and **J**, and is the only possible ground state.

7-31: The two $3s$ electrons have no orbital angular momentum, and their spins are aligned oppositely to give no net angular momentum. The $3p$ electron has $l = 1$, so **L** $= 1$, and in the ground state **J** $= \frac{1}{2}$. The term symbol is $^2P_{1/2}$.

7-33: A D state has **L** $= 2$; for a $2^2D_{3/2}$ state, $n = 2$ but **L** must always be strictly less than n, and so this state cannot exist.

7-35: (a) From Equation (7.17), $j = l \pm \frac{1}{2} = \frac{5}{2}, \frac{7}{2}$.

(b) Also from Equation (7.17), the corresponding angular momenta are $\frac{\sqrt{35}}{2}\hbar$ and $\frac{\sqrt{63}}{2}\hbar$.

(c) The values of L and S are $\sqrt{12}\,\hbar$ and $\frac{\sqrt{3}}{2}\hbar$. The law of cosines is

$$\cos\theta = \frac{J^2 - L^2 - S^2}{2LS},$$

where θ is the angle between **L** and **S**; then the angles are,

$$\arccos\left(\frac{(35/4) - 12 - (3/4)}{2\sqrt{12}\,((\sqrt{3})/2)}\right) = \arccos\left(-\frac{2}{3}\right) = 132.0°$$

and

$$\arccos\left(\frac{(63/4) - 12 - (3/4)}{2\sqrt{12}\,((\sqrt{3})/2)}\right) = \arccos\left(\frac{1}{2}\right) = 60.0°.$$

(d) The multiplicity is $2\left(\frac{1}{2}\right) + 1 = 2$, the state is an f state because the total angular momentum is provided by the f electron, and so the terms symbols are $^2F_{5/2}$ and $^2F_{7/2}$.

7-37: (a) In Figure 7.15, let the angle between **J** and **S** be α and the angle between **J** and **L** be β. Then, the product $\mu_J \hbar$ has magnitude

$$2\mu_B |\mathbf{S}| \cos\alpha + \mu_B |\mathbf{L}| \cos\beta = \mu_B |\mathbf{J}| + \mu_B |\mathbf{S}| \cos\alpha = \mu_B |\mathbf{J}| \left(1 + \frac{|\mathbf{S}|}{|\mathbf{J}|} \cos\alpha\right).$$

In the above, the factor of 2 in $2\mu_B$ relating the electron spin magnetic moment to the Bohr magneton is from Equation (7.3). The middle term is obtained by using $|\mathbf{S}|\cos\alpha + |\mathbf{L}|\cos\beta = |\mathbf{J}|$. The above expression is equal to the product $\mu_J \hbar$ because in this form, the magnitudes of the angular momenta include factors of $\hbar$.

From the law of cosines,

$$\cos\alpha = \frac{|\mathbf{L}|^2 - |\mathbf{J}|^2 - |\mathbf{S}|^2}{-2|\mathbf{J}||\mathbf{S}|},$$

and so

$$\frac{|\mathbf{S}|}{|\mathbf{J}|} \cos\alpha = \frac{|\mathbf{L}|^2 - |\mathbf{J}|^2 - |\mathbf{S}|^2}{2|\mathbf{J}|^2} = \frac{\mathbf{J}(\mathbf{J}+1) - \mathbf{L}(\mathbf{L}+1) + \mathbf{S}(\mathbf{S}+1)}{2\mathbf{J}(\mathbf{J}+1)},$$

and the expression for μ_J in terms of the quantum numbers is

$$\mu_J \hbar = |\mathbf{J}| g_J \mu_B, \qquad \text{or} \qquad \mu_J = \mathbf{J}(\mathbf{J}+1) g_J \mu_B, \qquad \text{where}$$

$$g_J = 1 + \frac{\mathbf{J}(\mathbf{J}+1) - \mathbf{L}(\mathbf{L}+1) + \mathbf{S}(\mathbf{S}+1)}{2\mathbf{J}(\mathbf{J}+1)}.$$

(b) There will be one substate for each value of M_J, where $M_J = -\mathbf{J} \ldots \mathbf{J}$, for a total of $2\mathbf{J}+1$ substates. The difference in energy between the substates is

$$\Delta E = g_J \mu_B M_J B.$$

7-39: The transitions that give rise to x-ray spectra are the same in all elements since the transitions involve only inner, closed-shell electrons. Optical spectra, however, depend upon the possible states of the outermost electrons, which, together with the transitions permitted for them, are different for atoms of different atomic number.

7-41: From either of Equations (7.21) or (7.22),

$$E = (10.2 \text{ eV})(Z-1)^2 = (10.2 \text{ eV})(144) = 1.47 \text{ keV}.$$

The wavelength is

$$\lambda = \frac{hc}{E} = \frac{1.240 \times 10^{-6} \text{ eV} \cdot \text{m}}{14.7 \times 10^3 \text{ eV}} = 8.44 \times 10^{-10} \text{ m} = 0.844 \text{ nm}.$$

7-43: In a singlet state, the spins of the outer electrons are antiparallel. In a triplet state, they are parallel.

Chapter 8 - Molecules

8-1: The nuclear charge of $+2e$ is concentrated at the nucleus, while the electron charges' densities are spread out in (presumably) the $1s$ subshell. This means that the additional attractive force of the two protons exceeds the mutual repulsion of the electrons to increase the binding energy.

8-3: Using 4.5 eV for the binding energy of hydrogen,

$$\frac{3}{2}kT = 4.5 \text{ eV} \quad \text{or} \quad T = \frac{2}{3}\frac{4.5 \text{ eV}}{8.617 \times 10^{-5} \text{ eV/K}} = 3.5 \times 10^4 \text{ K}.$$

8-5: The increase in bond lengths in the molecule increases its moment of inertia and accordingly decreases the frequencies in its rotational spectrum (see Equation (8.9)). In addition, the higher the quantum number J (and hence the greater the angular momentum), the faster the rotation and the greater the distortion, so the spectral lines are no longer evenly spaced.

Quantitatively, the parameter I (the moment of inertia of the molecule) is a function of J, with I larger for higher J. Thus, all of the levels as given by Equation (8.11) are different, so that the spectral lines are not evenly spaced. (It should be noted that if I depends on J, the algebraic steps that lead to Equation (8.11) will not be valid.)

8-7: From Equation (8.11), the ratios of the frequencies will be the ratio of the moments of inertia. For the different isotopes, the atomic separation, which depends on the charges of the atoms, will be essentially the same. The ratio of the moments of inertia will then be the ratio of the reduced masses. Denoting the unknown mass number by x and the ratio of the frequencies as r, r in terms of x is

$$r = \frac{\dfrac{x \cdot 16}{x + 16}}{\dfrac{12 \cdot 16}{12 + 16}}.$$

Solving for x in terms of r,

$$x = \frac{48\, r}{7 - 3\, r}.$$

Using $r = (1.153)/(1.102)$ in the above expression gives $x = 13.007$, or the integer 13 to three significant figures.

8-9: The corresponding frequencies are, from $\nu = \dfrac{c}{\lambda}$, and keeping an extra significant figure, in multiples of 10^{12} Hz:

$$2.484, \quad 3.113, \quad 4.337, \quad 4.947$$

The average spacing of these frequencies is $\Delta\nu = 0.616 \times 10^{12}$ Hz. (A least-squares fit from a spreadsheet program gives 0.6151 if $c = 2.998 \times 10^8$ m/s is used.) From Equation (8.11), the spacing of the frequencies should be $\Delta\nu = \dfrac{\hbar}{2\pi I}$; Solving for I and using $\Delta\nu$ as found above,

$$I = \frac{\hbar}{2\pi \Delta\nu} = \frac{(1.055 \times 10^{-34}\ \text{J·s})}{2\pi (0.6151 \times 10^{-12}\ \text{Hz})} = 2.73 \times 10^{-47}\ \text{kg·m}^2.$$

The reduced mass of the HCl molecule is $(35/36)m_H$, and so the distance between the nuclei is

$$R = \sqrt{\frac{I}{m'}} = \sqrt{\frac{(36)(2.73 \times 10^{-47}\ \text{kg·m}^2)}{(35)(1.6736 \times 10^{-27}\ \text{kg})}} = 0.129\ \text{nm}$$

(keeping extra significant figures in the intermediate calculation gives a result that is rounded to 0.130 nm to three significant figures).

8-11: Using $\nu_{1\to 0} = \dfrac{c}{\lambda}$ and $I = m' R^2$ in Equation (8.11) and solving for R,

$$R^2 = \frac{\hbar \lambda}{2\pi m' c}.$$

For this atom, $m' = m_H(200 \cdot 35)/(200 + 35)$, and

$$R = \sqrt{\frac{(1.055 \times 10^{-34}\ \text{J·s})(4.4 \times 10^{-2}\ \text{m})}{2\pi (1.6736 \times 10^{-27}\ \text{kg})(2.998 \times 10^8\ \text{m/s})} \frac{235}{200 \cdot 35}} = 0.223\ \text{nm},$$

or 0.22 nm to two significant figures.

8-13: Equation (8.11) may be re-expressed in terms of the frequency of the emitted photon when the molecule drops from the J rotational level to the $J-1$ rotational level,

$$\nu_{J \to J-1} = \frac{\hbar J}{2\pi I}.$$

For large J, the angular momentum of the molecule in its initial state is

$$L = \hbar\sqrt{J(J+1)} = \hbar J\sqrt{1 + \frac{1}{J}} \approx \hbar J.$$

Thus, for large J,

$$\nu \approx \frac{L}{2\pi I}, \qquad \text{or} \qquad L = \omega I,$$

the classical expression.

Molecules

8-15: The shape of the curve in Figure 8.18 will be the same for either isotope; that is, the value of k in Equation (8.14) will be the same. HD has the greater reduced mass, and hence the smaller frequency of vibration ν_0 and the smaller zero-point energy. HD is the more tightly bound, and has the greater binding energy since its zero-point energy contributes less energy to the splitting of the molecule.

8-17: (a) Using $m' = (19/20)m_H$ in Equation (8.15),

$$\nu_0 = \frac{1}{2\pi}\sqrt{\frac{966 \text{ N/m}}{(1.6736 \times 10^{-27} \text{ kg})}\frac{20}{19}} = 1.24 \times 10^{14} \text{ Hz}.$$

(b) $E_0 = (1/2)\hbar\sqrt{\dfrac{k}{m'}} = 4.11 \times 10^{-20}$ J. The levels are shown below, where the vertical scale is in units of 10^{-20} J and the horizontal scale is in units of 10^{-11} m.

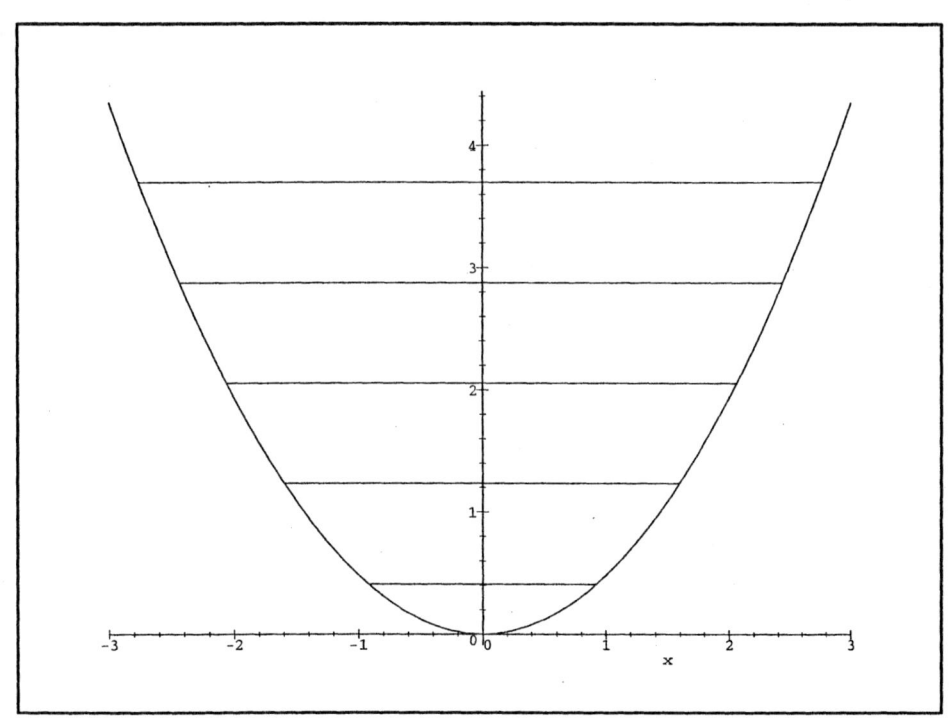

8-19: From Equation (8.16), the lower energy levels are separated by $\Delta E = h\nu_0$, and $\nu_0 = \Delta E/h$. Solving Equation (8.15) for k,

$$k = m'(2\pi\nu_0)^2 = m'\left(\frac{\Delta E}{\hbar}\right).$$

Using $m' = m_H(23 \cdot 35)/(23 + 35)$,

$$k = \frac{23 \cdot 35}{58}\left(1.6736 \times 10^{-27}\text{ kg}\right)\left(\frac{(0.063\text{ eV})\left(1.602 \times 10^{-19}\text{ J/eV}\right)}{(4.136 \times 10^{-15}\text{ eV·s})}\right) = 213\text{ N/m},$$

or 2.1×10^2 N/m to the given two significant figures.

8-21: Using

$$\Delta E = h\nu_0 = \hbar\sqrt{\frac{k}{m'}} \quad \text{and} \quad m' = m_H\frac{35}{36},$$

$$\Delta E = \left(1.055 \times 10^{-34}\text{ J·s}\right)\sqrt{\frac{516\text{ N/m}}{1.6736 \times 10^{-27}\text{ kg}}\frac{36}{35}} = 5.94 \times 10^{-20}\text{ J} = 0.371\text{ eV}.$$

At room temperature of about 300 K,

$$kT = \left(8.617 \times 10^{-5}\text{ eV/K}\right)(300\text{ K}) = 0.026\text{ eV}.$$

An individual atom is not likely to be vibrating in its first excited level, but in a large collection of atoms, it is likely that some of these atoms will be in the first excited state.

It's important to note that in the above calculations, the symbol "k" has been used for both a spring constant and Boltzmann's constant, quantities that are not interchangeable.

Chapter 9 - Statistical Mechanics

9-1: As in Example 9.1, $g(\epsilon_2) = 8$ and $g(\epsilon_1) = 1$. Then,

$$\frac{n(\epsilon_2)}{n(\epsilon_1)} = \frac{1}{1000} = 4e^{-(\epsilon_2 - \epsilon_1)/kT} = 4e^{3\epsilon_1/4kT},$$

where $\epsilon_2 = \epsilon_1/4$. Using $\epsilon_1 = -13.6$ eV, and solving for T,

$$T = \left(\frac{1}{k}\right)\frac{(3/4)(-\epsilon_1)}{\ln 4000} = \frac{(3/4)(13.6 \text{ eV})}{(8.617 \times 10^{-5} \text{ eV/K})(\ln 4000)} = 1.43 \times 10^4 \text{ K}.$$

9-3: In this situation, the "multiplicity" that is the presuperscript in the term symbol is not the same as the number of states of a given energy. The number of states is $2\mathbf{L} + 1 = 3$ for a P level and 1 for an S level. The ratio of the numbers of atoms in the states is then

$$(3)\exp\left(-\frac{(2.093 \text{ eV})}{(8.617 \times 10^{-5} \text{ eV/K})(1200 \text{ K})}\right) = 4.86 \times 10^{-9}.$$

9-5: (a) As in Example 9.2, there are $2J + 1$ states with the same rotational energy for a given rotational quantum number J. The $J = 0$ state has 0 energy, and so the populations relative to $J = 0$ are

$$(2J+1)\exp\left(-\frac{J(J+1)\hbar^2}{2IkT}\right)$$

$$= (2J+1)\left[\exp\left(-\frac{\hbar^2}{2IkT}\right)\right]^{J(J+1)}$$

$$= (2J+1)\left[\exp\left(-\frac{(1.055 \times 10^{-34} \text{ J·s})^2}{2(4.64 \times 10^{-48} \text{ kg·m}^2)(1.381 \times 10^{-23} \text{ J/K})(300 \text{ K})}\right)\right]^{J(J+1)}$$

$$= (2J+1)[0.74864]^{J(J+1)}.$$

Applying this expression to $J = 0, 1, 2, 3$ and 4 gives, respectively, 1 exactly, 1.68, 0.880, 0.217 and 0.275.

If more precise values for the constants $\hbar$ and k than those given in the endpapers are used, the answers might differ in the third figure. For instance, using $\hbar = 1.0545716 \times 10^{-34}$ J·s gives, for $J = 2, 3$ and 4 the values 0.882, 0.218 and 0.277.

(b) Introduce the dimensionless parameter $x \equiv e^{-\hbar^2/2IkT}$ (in part (a), with $T = 300$ K, $x = 0.74864$). Then, for the populations of the $J = 2$ and $J = 3$ states to be equal,

$$5x^6 = 7x^{12}, \qquad x^6 = \frac{5}{7} \qquad \text{and} \qquad 6 \ln x = \ln \frac{5}{7}.$$

Using $\ln x = -\hbar^2/2IkT$ and $\ln \frac{5}{7} = -\ln \frac{7}{5}$, and solving for T,

$$T = \frac{6\hbar^2}{2Ik \ln(7/5)}$$

$$= \frac{6\left(1.055 \times 10^{-34} \text{ J·s}\right)^2}{2\left(4.64 \times 10^{-48} \text{ kg·m}^2\right)\left(1.381 \times 10^{-23} \text{ J/K}\right) \ln(1.4)} = 1.55 \times 10^3 \text{ K}.$$

9-7: The mean speed $\overline{v} = \frac{1}{2}(1.00 \text{ m/s} + 3.00 \text{ m/s}) = 2.00$ m/s. The root-mean-square speed is

$$v_{\text{rms}} = \sqrt{\frac{1}{2}\left((1.00 \text{ m/s})^2 + (2.00 \text{ m/s})^2\right)} = 2.24 \text{ m/s}.$$

9-9: For monatomic hydrogen, the kinetic energy is all translational and $\overline{\text{KE}} = (3/2)kT$; solving for T with $\overline{\text{KE}} = -E_1$,

$$T = \frac{2}{3}\left(-\frac{E_1}{k}\right) = \frac{(2/3)(13.6 \text{ eV})}{(8.617 \times 10^{-5} \text{ eV/K})} = 1.05 \times 10^5 \text{ K}.$$

9-11: For these nonrelativistic atoms, the shift in wavelengths will be between $+\lambda(v/c)$ and $-\lambda(v/c)$ and the width of the doppler-broadened line will be $2\lambda(v/c)$. Using the rms speed from $\overline{\text{KE}} = (3/2)kT = (1/2)mv^2$, $v = \sqrt{3kT/m}$, and

$$\Delta \lambda = 2\lambda \frac{\sqrt{3kT/m}}{c}$$

$$= 2\left(656.3 \times 10^{-9} \text{ m}\right) \frac{\sqrt{3\left(1.381 \times 10^{-23} \text{ J/K}\right)(500 \text{ K})/\left(1.6736 \times 10^{-27} \text{ kg}\right)}}{(2.998 \times 10^8 \text{ m/s})}$$

$$= 1.54 \times 10^{-11} \text{ m} = 15.4 \text{ pm}.$$

Statistical Mechanics

9-13: The average value of $\dfrac{1}{v}$ is

$$\left\langle \frac{1}{v} \right\rangle = \frac{1}{N} \int_0^\infty \frac{1}{v} n(v)\, dv.$$

With $n(v)$ as given in Equation (9.14),

$$\left\langle \frac{1}{v} \right\rangle = \frac{1}{N} 4\pi N \left(\frac{m}{2\pi kT}\right)^{3/2} \int_0^\infty v\, e^{-mv^2/2kT}\, dv$$

$$= 4\pi \left(\frac{m}{2\pi kT}\right)^{3/2} \left(\frac{kT}{m}\right) = \sqrt{\frac{2m}{\pi kT}},$$

where the improper definite integral given in the problem,

$$\int_0^\infty v\, e^{-av^2}\, dv = \frac{1}{2a}$$

has been used.

Note that
$$\left\langle \frac{1}{v} \right\rangle = 2 \frac{1}{\langle v \rangle},$$

where the notation $\bar{v} = \langle v \rangle$ has been used.

9-15: See Figure 9.5. The curves are not normalized, in that when α is adjusted to give the same areas under the curves, the curves will interect at a finite energy. A fermion gas will exert the greatest pressure because the Fermi distribution has a larger proportion of high-energy particles than the other distributions (note that the *proportion* of high-energy particles is larger). A boson gas will exert the least pressure because the Bose distribution has a larger proportion of low-energy particles than the others.

9-17: NOTE: For convenience in this problem, the quantity $g(\lambda)$ is used as well as $g(\nu)$, even though they are different functions, with different arguments and, as will be seen, different functional forms.

The condition that $g(\lambda)$ must satisfy is

$$g(\nu)\, d\nu = g(\lambda)\, d\lambda, \quad \text{so} \quad g(\lambda) = g(\nu) \frac{d\nu}{d\lambda}.$$

The quantity $\dfrac{d\nu}{d\lambda}$ is negative, so it is convenient and conventional to use instead

$$g(\lambda) = g(\nu) \left|\frac{d\nu}{d\lambda}\right| = g(\nu) \frac{c}{\lambda^2} = \frac{8\pi L^3 \nu^2}{c^2 \lambda^2} = \frac{8\pi L^3}{\lambda^4},$$

where Equation (9.34) has been used for $g(\nu)$.

The number of waves between 9.5 mm and 10.5 mm is then

$$g(\lambda)\Delta\lambda = \frac{8\pi (1\text{ m})^3}{(10\text{ mm})^4}(1.00\text{ mm}) = 2.5\times 10^6.$$

(It should be noted that integrating $g(\lambda)\,d\lambda$ between the frequencies and including the variation of $g(\lambda)$ with λ, instead of using $g(\lambda) = g(\overline{\lambda})$, as was done above, gives the same answer to the given precision.)

Similarly, the number of waves between 99.5 mm and 100.5 mm is 2.5×10^2, lower by a factor of 10^4.

9-19: The percentage difference is the percentage difference in the fourth powers of the Kelvin temperatures; specifically,

$$\frac{\sigma T_1^4 - \sigma T_2^4}{\sigma T_1^4} = \frac{T_1^4 - T_2^4}{T_1^4} = 1 - \left(\frac{T_2}{T_1}\right)^4 = 1 - \left(\frac{307\text{ K}}{308\text{ K}}\right)^4 = 0.013 = 1.3\%.$$

For temperature variations this small, the fractional variation may be approximated by

$$\frac{\Delta(T^4)}{T^4} = \frac{3T^3\,\Delta T}{T^4} = 3\frac{\Delta T}{T} = 3\frac{1\text{ K}}{308\text{ K}} = 0.013$$

to the given precision.

9-21: See Example 9.7. Lowering the Kelvin temperature by a given fraction will lower the radiation by a factor equal to the fourth power of the ratio of the temperatures. Using 1.4 kW/m² as the rate at which the sun's energy arrives at the surface of the earth,

$$(1.4\text{ kW/m}^2)(0.90)^4 = 0.92\text{ kW/m}^2.$$

9-23: To radiate at twice the rate, the fourth power of the Kelvin temperature would need to double. The new temperature would be

$$((400+273)\text{ K})\,2^{1/4} = 800\text{ K},$$

which is 527°C.

9-25: From Equation (9.41), with unit emissivity for the hole in the wall,

$$P = \sigma T^4 = (5.670\times 10^{-8}\text{ W}/(\text{m}^2\cdot\text{K}^4))(973\text{ K})^4(10\times 10^{-4}\text{ m}^2) = 51\text{ W}.$$

9-27: Using Equation (9.41) for the radiated power per unit area, the area of the blackbody (assuming unit emissivity) is

$$A = \frac{P}{R} = \frac{P}{e\sigma T^4} = \frac{(1.00 \times 10^3 \text{ W})}{(1)(5.670 \times 10^{-8} \text{ W}/(\text{m}^2 \cdot \text{K}^4))((500+273) \text{ K})^4}$$
$$= 4.94 \times 10^{-2} \text{ m}^2 = 494 \text{ cm}^2.$$

The radius of a sphere with this surface area is found from $A = 4\pi r^2$, or

$$r = \sqrt{\frac{A}{4\pi}} = \sqrt{\frac{494 \text{ cm}^2}{4\pi}} = 6.27 \text{ cm}.$$

9-29: Equation (9.38) is not integrable in terms of elementary functions; however, approximating $g(\nu)$ by $g(\overline{\nu})$, where $\overline{\nu}$ is the average frequency in the wavelength interval ($\overline{\nu}$ will be approximated by $c/\overline{\lambda}$), is valid. Before using Equation (9.38), it is convenient to note that because the proportion of the radiation in this wavelength interval is desired, division by $u = \int u(\nu)\,d\nu$ gives the fraction

$$\frac{\Delta u}{u} = 15 \left(\frac{h\nu}{\pi kT}\right)^4 \frac{(\Delta \nu/\nu)}{e^{h\nu/kT}-1}.$$

The quantity $\Delta\nu/\nu$ is approximated by $\Delta\lambda/\lambda$ (supressing the minus sign). The dimensionless quantity $h\nu/kT$ that appears in the above expression is evaluated at the average frequency, in terms of the average wavelength, as outlined above, so that

$$\frac{h\overline{\nu}}{kT} = \frac{hc}{kT\overline{\lambda}} = \frac{(1.240 \times 10^{-6} \text{ eV} \cdot \text{m})}{(8.617 \times 10^{-5} \text{ eV/K})(6000 \text{ K})(580 \times 10^{-9} \text{ m})} = 4.135,$$

keeping extra significant figures. The result is

$$\frac{\Delta u}{u} = 15 \left(\frac{4.135}{\pi}\right)^4 \frac{(20 \text{ nm})/(580 \text{ nm})}{\exp(4.135)-1} = 0.025 = 2.5\%.$$

To do the integral numerically, an example of a sequence of Maple commands that reproduce the above result, and allows similar calculations for arbitrary ranges of wavelengths and temperatures, is:

```
>u:=x^3/(exp(y*x)-1);
>y:=1.24E-6/8.167E-5/6E3;
>x1:=1/590E-9; x2:=1/570E-9;
>evalf(int(u,x=x1..x2))/int(u,x=0..infinity);
```

9-31: From the Wien displacement law (Equation (9.40)), the surface temperature of Sirius is

$$T = \frac{2.898 \times 10^{-3} \text{ m·K}}{290 \times 10^{-9} \text{ m}} = 1.0 \times 10^4 \text{ K}.$$

9-33: From the Wien displacement law (Equation (9.40)), the surface temperature of the cloud is

$$T = \frac{2.898 \times 10^{-3} \text{ m·K}}{10 \times 10^{-6} \text{ m}} = 2.9 \times 10^2 \text{ K}$$

(the form of the answer indicates that this result is valid to no more than two significant figures).

Assuming unit emissivity, the radiation rate is

$$R = \sigma T^4 = \frac{P}{A} = \frac{P}{\pi D^2},$$

where D is the cloud's diameter. Solving for D using the given power and the temperature found above,

$$D = \sqrt{\frac{P}{\pi \sigma T^4}} = 8.9 \times 10^{11} \text{ m},$$

roughly but slightly larger than the distance from the sun to Jupiter.

9-35: The total energy (denoted by uppercase U) is related to the energy density by $U = V u$, where V is the volume. In terms of the temperature,

$$U = V u = V a T^4 = V \frac{4\sigma}{c} T^4.$$

The specific heat at constant volume is then

$$\begin{aligned}
c_V = \frac{\partial U}{\partial T} &= \frac{16\sigma}{c} T^3 V \\
&= \frac{16 \left(5.670 \times 10^{-8} \text{ W/}\left(\text{m}^2\cdot\text{K}^4\right)\right)}{(2.998 \times 10^8 \text{ m/s})} (1000 \text{ K})^3 \left(1.00 \times 10^{-6} \text{ m}^3\right) \\
&= 3.03 \times 10^{-12} \text{ J/K}.
\end{aligned}$$

Statistical Mechanics

9-37: At $T = 0$, all states with energy less than the Fermi energy ϵ_F are occupied, and all states with energy above the Fermi energy are empty. For $0 \leq \epsilon \leq \epsilon_F$, the electron energy distribution, given in Equation (9.58), is proportional to $\sqrt{\epsilon}$. The median energy is that energy for which there are as many occupied states below the median as there are above. The median energy ϵ_M is then the energy such that

$$\int_0^{\epsilon_M} \sqrt{\epsilon}\, d\epsilon = \frac{1}{2} \int_0^{\epsilon_F} \sqrt{\epsilon}\, d\epsilon.$$

Evaluating the integrals,

$$\frac{2}{3}(\epsilon_M)^{3/2} = \frac{1}{3}(\epsilon_F)^{3/2}, \quad \text{or} \quad \epsilon_M = \left(\frac{1}{2}\right)^{3/2} \epsilon_F = 0.630\, \epsilon_F.$$

9-39: (a) The average energy at $T = 0$, from Equation (9.59), is $(3/5)\epsilon_F = 3.31$ eV.

(b) Setting $(3/2)kT = (3/5)\epsilon_F$ and solving for T,

$$T = \frac{2}{5}\frac{\epsilon_F}{k} = \frac{2}{5}\frac{5.51 \text{ eV}}{8.617 \times 10^{-5} \text{ eV/K}} = 2.56 \times 10^4 \text{ K}.$$

(c) The speed in terms of the kinetic energy is

$$v = \sqrt{\frac{2\,\text{KE}}{m}} = \sqrt{\frac{6\,\epsilon_F}{5m}} = \sqrt{\frac{6\,(5.51 \text{ eV})(1.602 \times 10^{-19} \text{ J/eV})}{5\,(9.1095 \times 10^{-31} \text{ kg})}} = 1.08 \times 10^6 \text{ m/s}.$$

9-41: The denominator is not well-defined at $T = 0$, but the expression in Equation (9.29) may be evaluated by taking the limit as $T \to 0^+$. If $\epsilon > \epsilon_F$, the argument of the exponent is positive for positive T, and as $T \to 0^+$ the exponent becomes unboundedly large and $f_{FD}(\epsilon) \to 0$. If $\epsilon < \epsilon_F$, the argument of the exponent is always negative and the exponent goes to zero as $T \to 0^+$, so the denominator approaches 1 and $f_{FD} \to 1$. The interpretation of these results is that in the limit $T \to 0^+$, states with $\epsilon > \epsilon_F$ are unoccupied and states with $\epsilon < \epsilon_F$ are fully occupied.

9-43: Using Equation (9.29),

$$f_1 = f_{FD}(\epsilon_F + \Delta\epsilon) = \frac{1}{e^{\Delta\epsilon/kT} + 1}, \quad \text{and}$$

$$f_2 = f_{FD}(\epsilon_F + \Delta\epsilon) = \frac{1}{e^{-\Delta\epsilon/kT} + 1}.$$

From these expressions,

$$f_1 + f_2 = \frac{1}{e^{\Delta\epsilon/kT}+1} + \frac{1}{e^{-\Delta\epsilon/kT}+1}$$
$$= \frac{1}{e^{\Delta\epsilon/kT}+1} + \frac{e^{\Delta\epsilon/kT}}{e^{\Delta\epsilon/kT}+1}$$
$$= 1.$$

9-45: In using Equation (9.56) to find the Fermi energy, the proper values for N/V, the number of free electrons per unit volume, and m^*, the effective electron mass, must be used. From Table 7.4, zinc in its ground state has two electrons in the $4s$ subshell and completely filled K, L and M shells. Thus, there are two free electrons per atom. The number of atoms per unit volume is the ratio of the mass density ρ_{Zn} to the mass per atom m_{Zn}. Combining in Equation (9.56),

$$\epsilon_F = \frac{h^2}{2\,m^*}\left(\frac{3\,(2)\,\rho_{Zn}}{8\pi\,m_{Zn}}\right)^{2/3}$$
$$= \left(\frac{(6.626\times 10^{-34}\ \text{J·s})^2}{2\,(0.85)\,(9.1095\times 10^{-31}\ \text{kg})}\right)\left(\frac{3\,(2)\,(7.13\times 10^3\ \text{kg/m}^3)}{8\pi\,(65.4\ \text{u})\,(1.66054\times 10^{-27}\ \text{kg/u})}\right)^{2/3}$$
$$= 1.78\times 10^{-18}\ \text{J} = 11\ \text{eV}$$

to two significant figures.

9-47: At $T=0$, the electron distribution $n(\epsilon)$ as given in Equation (9.58) reduces to

$$n(\epsilon) = \frac{3\,N}{2}\,(\epsilon_F)^{-3/2}\,\sqrt{\epsilon},$$

as explained in the derivation of Equation (9.59) and in Problem 9-41.

At $\epsilon = (\epsilon_F)/2$,

$$n\left(\frac{\epsilon_F}{2}\right) = \frac{3}{\sqrt{8}}\,\frac{N}{\epsilon_F}.$$

The number of atoms is the mass divided by the mass per atom,

$$N = \frac{(1.00\times 10^{-3}\ \text{kg})}{(63.55\ \text{u})\,(1.66054\times 10^{-27}\ \text{kg/u})} = 9.48\times 10^{21},$$

with the atomic mass of copper from the front endpapers and $\epsilon_F = 7.04$ eV is from Table 9.2 or Problem 9-40. The number of states per electronvolt is

$$n\left(\frac{\epsilon_F}{2}\right) = \frac{3}{\sqrt{8}}\,\frac{9.48\times 10^{21}}{7.04\ \text{eV}} = 1.43\times 10^{21}\ \text{states/eV},$$

and the distribution may certainly be considered to be continuous.

Statistical Mechanics 79

9-49: Using the approximation $f(\epsilon) = A e^{-\epsilon/kT}$, and a factor of 4 instead of 8 in Equation (9.47), Equation (9.57) becomes

$$n(\epsilon)\, d\epsilon = g(\epsilon)\, f(\epsilon)\, d\epsilon = A\, 4\sqrt{2}\, \pi\, \frac{V\, m^{3/2}}{h^3}\, \sqrt{\epsilon}\, e^{-\epsilon/kT}\, d\epsilon.$$

Integrating over all energies,

$$N = \int_0^\infty n(\epsilon)\, d\epsilon = A\, 4\sqrt{2}\, \pi\, \frac{V\, m^{3/2}}{h^3} \int_0^\infty \sqrt{\epsilon}\, e^{-\epsilon/kT}\, d\epsilon.$$

The integral is that given in the problem with $x = \epsilon$ and $a = kT$,

$$\int_0^\infty \sqrt{\epsilon}\, e^{-\epsilon/kT}\, d\epsilon = \frac{\sqrt{\pi\, kT}}{2\, (1/kT)} = \frac{\sqrt{\pi\, (kT)^3}}{2}, \quad \text{so that}$$

$$N = A\, 4\sqrt{2}\, \pi\, \frac{V\, m^{3/2}}{h^3} \frac{\sqrt{\pi\, (kT)^3}}{2} = A\, \frac{V}{h^3}\, (2\pi\, m\, kT)^{3/2}.$$

Solving for A,

$$A = \frac{N}{V}\, h^3\, (2\pi\, m\, kT)^{-3/2}.$$

Using the given numerical values,

$$A = \left(\frac{\frac{6.022 \times 10^{26}\ \text{kmol}^{-1}}{22.4\ \text{kg/kmol}} \left(6.626 \times 10^{-34}\ \text{J·s}\right)^3 \times}{\left[2\pi\, (4.00\ \text{u})\, (1.66054 \times 10^{-27}\ \text{kg/u})\, (1.381 \times 10^{-23}\ \text{J/K})\, (293\ \text{K})\right]^{-3/2}} \right)$$

$$= 3.56 \times 10^{-6},$$

which is much less than one. In the above calculation, care must be taken in evaluating A; in SI units, the exponent of h^3 is greater than 100, and will cause difficulty if tried on some standard calculators. A possible method of evaluation is to find the last term first, multiply by h and then multiply by h^2.

9-51: See Problem 9-49. Here, the original factor of 8 must be retained, with the result that

$$A = \frac{1}{2} \frac{N}{V}\, h^3\, (2\pi\, m_e\, kT)^{-3/2}$$

$$= \left(\begin{array}{c} (1/2)\, (8.48 \times 10^{28}\ \text{m}^{-3})\, (6.626 \times 10^{-34}\ \text{J·s})^3 \times \\ \left[2\pi\, (9.1095 \times 10^{-31}\ \text{kg})\, (1.381 \times 10^{-23}\ \text{J/K})\, (293\ \text{K})\right]^{-3/2} \end{array} \right)$$

$$= 3.50 \times 10^3,$$

which is much greater than one, and so the Fermi-Dirac distribution cannot be approximated by a Maxwell-Boltzmann distribution. (See the above note for Problem 9-49 regarding the difficulties involved in using h^3 numerically.)

9-53: The number density (N_{atom}/V) for either gas is the ratio of the total mass and the mass of a single atom, divided by the volume (assumed spherical);

$$\frac{N}{V} = \frac{M_{star}}{m_C} \frac{1}{(4\pi/3) R_{star}^3}$$

$$= \frac{(2.0 \times 10^{30} \text{ kg})/2}{(12 \text{ u})(1.66054 \times 10^{-27} \text{ kg/u})} \frac{1}{(4\pi/3)((0.010) \times 7.0 \times 10^8 \text{ m})^3}$$

$$= 3.49 \times 10^{34} \text{ atoms/m}^3.$$

(a) The Fermi energy of the carbon nucleus gas is found with the above value of (N_{atom}/V), with, of course, one nucleus per atom;

$$\epsilon_F = \frac{h^2}{2 m_C} \left(\frac{3}{8\pi} \frac{N_{atom}}{V} \right)^{2/3}$$

$$= \frac{(6.626 \times 10^{-34} \text{ J·s})^2}{2(12 \text{ u})(1.66054 \times 10^{-27} \text{ kg/u})} \left(\frac{3}{8\pi} (3.49 \times 10^{34} \text{ nuceli/m}^3) \right)^{2/3}$$

$$= 2.85 \times 10^{-19} \text{ J} = 1.78 \text{ eV}.$$

Because there are six electrons per carbon atom, the Fermi energy of the electron gas is found with 6 times the above value of (N_{atom}/V);

$$\epsilon_F = \frac{h^2}{2 m_e} \left(\frac{3}{8\pi} \frac{6 N_{atom}}{V} \right)^{2/3}$$

$$= \frac{(6.626 \times 10^{-34} \text{ J·s})^2}{2(9.1095 \times 10^{-31} \text{ kg})} \left(\frac{18}{8\pi} (3.49 \times 10^{34} \text{ nuceli/m}^3) \right)^{2/3}$$

$$= 2.06 \times 10^{-14} \text{ J} = 129 \text{ keV}$$

(use of less precise values for the constants, or roundoff in intermediate calculations, may lead to a result that differs in the last significant figure).

(b) For either gas, $kT = (8.617 \times 10^{-5} \text{ eV/K})(10^7 \text{ K}) = 862$ eV. The gas of nuclei is nondegenerate, and the gas of electrons will be mostly degenerate (the factor $kT/\epsilon_F \approx 10^{-2}$, and there will be a small fraction of the electrons above the Fermi energy).

Chapter 10 - The Solid State

10-1: The halogenic atomic numbers are $Z=9$ for flourine (F), $Z=17$ for chlorine (Cl), $Z=35$ for bromine (Br) and $Z=53$ for iodine (I). The greater the atomic number of a halogen atom, the larger the atom is, hence the increase in the interatomic spacing with Z. The larger the ion spacing, the smaller the cohesive energy, hence the lower the melting point.

10-3: (a) The cohesive energy will be the negative of the Coulombic energy as given in Equation (10.1), minus the difference between the ionization energy of potassium and the electron affinity of chlorine;

$$\alpha \frac{e^2}{4\pi \epsilon_0 r} - (E_i - E_a)$$
$$= (1.748)\frac{(8.988 \times 10^9 \text{ N·m}^2/\text{C}^2)(1.602 \times 10^{-19} \text{ C})^2}{(0.314 \times 10^{-9} \text{ m})(1.602 \times 10^{-19} \text{ J/eV})} - (4.34 \text{ eV} - 3.61 \text{ eV})$$
$$= 7.29 \text{ eV}.$$

(b) The difference between the observed binding energy and that found in part (a) must be due to the repulsive energy as given in Equation (10.1). From the observed binding energy, U_0 must be given by

$$-U_0 = 6.42 \text{ eV} + (4.34 \text{ eV} - 3.61 \text{ eV}) = 7.15 \text{ eV}.$$

The Coulombic energy, an intermediate calculation in part (a), is $U_{\text{coulomb}} = -8.0156$ eV, and so solving Equation (10.5) for n,

$$n = \left[1 - \frac{U_0}{U_{\text{coulomb}}}\right]^{-1} = \left[1 - \frac{7.15 \text{ eV}}{8.0156 \text{ eV}}\right]^{-1} = 9.26.$$

10-5: The heat lost by the expanding gas is equal to the work done against the attractive van der Waals forces between the gas molecules.

10-6: Van der Waals forces are too weak to hold inert gas atoms together against the forces exerted during collisions in the gaseous state.

10-7: (a) Van der Waals forces increase the cohesive energy because they are attractive, and the ions in the crystals tend more to cohere. (b) Zero-point oscillations decrease the cohesive energy because these oscillations represent a mode of energy that is present in a solid but not in individual atoms or ions.

10-9: The electrons that constitute the "gas" of freely moving electrons are only those that are loosely bound to the nuclei, specifically those electrons in the outer shells. As has been seen, the innermost electrons have binding energies that give rise to x-ray spectra, and will not be members of the free-electron gas.

10-11: The number density n is the mass density ρ_{Ag} divided by the mass m_{Ag} of each atom,

$$n = \frac{\rho_{\text{Ag}}}{m_{\text{Ag}}} = \frac{(10.5 \times 10^3 \text{ kg/m}^3)}{(108 \text{ u})(1.66054 \times 10^{-27} \text{ kg/u})} = 5.855 \times 10^{28} \text{ atoms/m}^3,$$

keeping an extra significant figure. With the assumption stated in the problem, this is the same as the electron density.

Using $\lambda = 200\, d = 200\, n^{-1/3}$ and $m\, v_F = \sqrt{2\, m\, \epsilon_F}$ in Equation (10.16),

$$\rho = \frac{m_e\, v_F}{n\, d^2\, (200\, n^{-1/3})} = \frac{\sqrt{2\, m_e\, \epsilon_F}}{200\, e^2\, n^{2/3}}$$

$$= \frac{\sqrt{2\,(9.1095 \times 10^{-31} \text{ kg})(5.51 \text{ eV})(1.602 \times 10^{-19} \text{ J/eV})}}{(200)(1.602 \times 10^{-19} \text{ C})^2 (5.855 \times 10^{28} \text{ m}^{-3})^{2/3}} = 1.64 \times 10^{-8}\ \Omega\cdot\text{m}$$

10-13: In both insulators and semiconductors, a forbidden band separates a filled valence band from the conduction band above it. In semiconductors, the band gap is smaller than in insulators, and the property of the gap that makes a semiconductor a semiconductor is that in the semiconductor some valence electrons have enough thermal energy to jump across the gap to the conduction band.

10-15: (a) Photons of visible light have energies of $\sim$ 1-3 eV (see the back endpapers), which can be absorbed by free electrons in a metal without leaving the electrons' valence band. Hence metals are opaque. The forbidden bands in insulators and semiconductors are too wide for valence electrons to jump across the gaps by absorbing 1-3 eV. Hence such solids are transparent.

(b) The minimum wavelengths are given by $\lambda = hc/E_g$, where E_g is the gap energy; light with shorter wavelengths would have energies larger than the gap energy. For silicon,

$$\lambda_{\min} = \frac{hc}{E_g} = \frac{1.240 \times 10^{-6} \text{ eV}\cdot\text{m}}{1.1 \text{ eV}} = 1.13 \times 10^{-6} \text{ m} = 1.13\ \mu\text{m} = 1130 \text{ nm},$$

The Solid State

keeping an extra significant figure; elemental silicon is not transparent to visible light. For diamond,

$$\lambda_{\min} = \frac{hc}{E_g} = \frac{1.240 \times 10^{-6} \text{ eV} \cdot \text{m}}{6 \text{ eV}} = 2.07 \times 10^{-7} \text{ m} = 207 \text{ nm},$$

again keeping extra significant figures. Light with this wavelength is in the ultraviolet, but light with *longer* wavelengths, including the visible wavelengths, have less energy and cannot excite an electron to jump the band, and will pass through. Hence, diamond is transparent to visible light.

10-17: Aluminum atoms have 3 electrons in their outer shell, germanium atoms have 4 (see Table 7.4). Replacing a germanium atom with an aluminum atom leaves a hole, and the the result is a *p*-type semiconductor.

10-19: The figure on the next page shows the third Brillouin zone. The wavevectors in this zone will be those that do not fit into the first two zones but are not diffracted by the diagonal sets of atomic planes in Figure 10.40 that make angles of $\pm \arctan(1/2) = \pm 26.6°$ or $\pm \arctan(2) = \pm 63.4°$ with the *x*- or *y*-axes. These wavenumbers correspond to $|k_x| + |k_y| > \dfrac{2\pi}{a}$ (the condition that **k** not be in the first or second zones) and

$$\left\{ \frac{\pi}{a} < |k_x| < \frac{2\pi}{a},\ 0 < |k_y| > \frac{\pi}{a} \right\} \quad \text{OR} \quad \left\{ 0 < |k_x| < \frac{\pi}{a},\ \frac{\pi}{a} < |k_y| > \frac{2\pi}{a} \right\}$$

Figure for Solution to Problem 9-19: More text on previous page

In the figure, the darker-shaded square is the first Brillouin zone and the lighter-shaded square is the second zone, as in Figure 10.41 (the axes and axis scales are not shown, but are the same as in the text figure). The unshaded area is the third Billouin zone.

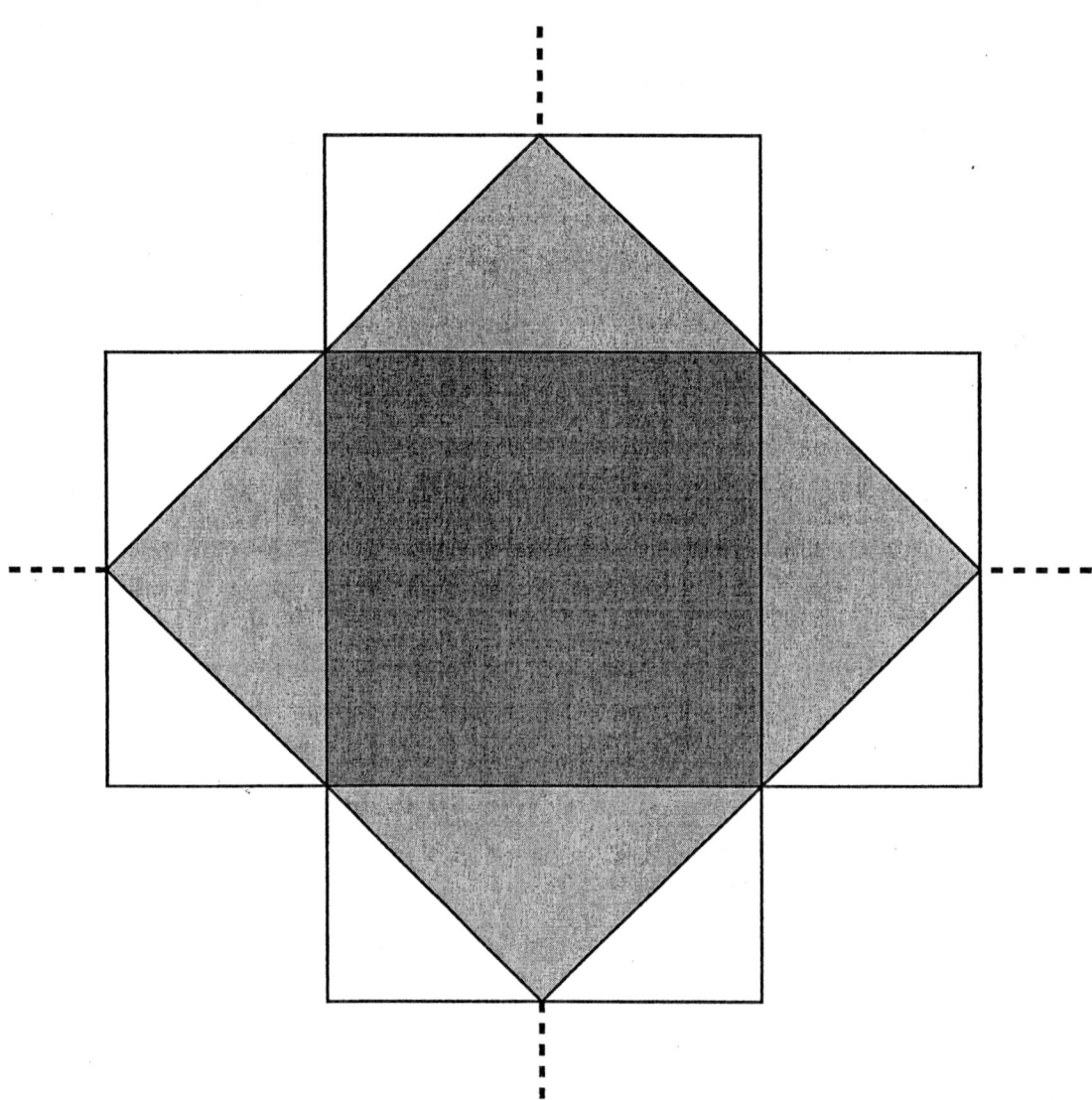

10-21: (a) The radius of a Bohr orbit for a given energy level n is proportional to the relative permittivity (the dielectric constant) and the mass, as given in Equation (4.13). The radius of the first Bohr orbit, in terms of a_0, would then be

$$r_1 = a_0 \frac{(\epsilon/\epsilon_0)}{(m^*/m)} = \left(5.292 \times 10^{-11} \text{ m}\right) \frac{16}{0.17} = 5.0 \text{ nm}.$$

The Solid State   85

(b) The energy of an electron in a given Bohr orbit is porportional to the effective mass and inversely proportional to the square of the relative permittivity (see Equation (4.15)). The ionization energy is then

$$E = (-E_1) \frac{(m^*/m)}{(\epsilon/\epsilon_0)^2} = (13.6 \text{ eV}) \frac{(0.17)}{(16)^2} = 9.0 \times 10^{-3} \text{ eV}.$$

This is much smaller than the energy gap of 0.65 eV but comparable to the product $kT \approx 0.025$ eV.

10-23: (a) In a uniform magnetic field of magnitude B, an electron moving with nonrelativistic speed v at right angles to the field will experience a force of magntiude evB. The inward acceleration of a particle moving with constant speed v in a circle of radius R has magnitude v^2/R. From $F = m^*a$,

$$evB = m^* \frac{v^2}{R}, \quad \text{or} \quad eB = m^* (2\pi \nu_c),$$

where $\nu_c = v/(2\pi R)$ has been used. Solving for the cyclotron frequency ν_c,

$$\nu_c = \frac{eB}{2\pi m^*}.$$

Note that the cyclotron frequency is independent of the electron's speed, and hence its orbit radius.

(b) From the above expression for ν_c, solving for m^* gives

$$m^* = \frac{eB}{2\pi \nu_c} = \frac{(1.602 \times 10^{-19} \text{ C})(0.1 \text{ T})}{2\pi (1.4 \times 10^{10} \text{ Hz})} = 1.82 \times 10^{-31} \text{ kg} = 0.2\, m_e.$$

(c) Solving for R as a function of v and ν,

$$R = \frac{v}{2\pi \nu} = \frac{(3 \times 10^4 \text{ m/s})}{2\pi (1.4 \times 10^{10} \text{ Hz})} = 3.4 \times 10^{-7} \text{ m}.$$

10-25: From Equation (10.28), the frequency is

$$\nu = \frac{2(Ve)}{h} = \frac{2(5.0 \times 10^{-6} \text{ eV})}{4.136 \times 10^{-15} \text{ eV} \cdot \text{s}} = 2.4 \times 10^9 \text{ Hz} = 2.4 \text{ GHz}.$$

Chapter 11 - Nuclear Structure

11-1: $^{6}_{3}\text{Li}$: $Z = 3$ protons, $A - Z = 6 - 3 = 3$ neutrons.

$^{22}_{10}\text{Ne}$: $Z = 10$ protons, $A - Z = 22 - 10 = 12$ neutrons.

$^{94}_{40}\text{Zr}$: $Z = 40$ protons, $A - Z = 94 - 40 = 54$ neutrons.

$^{180}_{72}\text{Hf}$: $Z = 72$ protons, $A - Z = 180 - 72 = 108$ neutrons.

11-3: The radius of a gold nucleus is, from Equation (11.1),

$$R = R_0 A^{1/3} = (1.2 \times 10^{-15} \text{ m})(197)^{1/3} = 6.98 \times 10^{-15} \text{ m}.$$

The momentum of an electron with this wavelength is $p = h/\lambda$, and the kinetic energy is

$$\text{KE} = E - mc^2 = \sqrt{(pc)^2 + (mc^2)^2} - mc^2 = \sqrt{\left(\frac{hc}{\lambda}\right)^2 + (mc^2)^2} - mc^2$$

$$= \sqrt{\left(\frac{1.240 \times 10^{-6} \text{ eV}\cdot\text{m}}{6.98 \times 10^{-15} \text{ m}}\right)^2 + (0.511 \text{ MeV})^2} - (0.511 \text{ MeV})$$

$$= 177 \text{ MeV}.$$

11-5: From Equation (11.1), the radius of such a nucleus would be

$$R = R_0 A^{1/3} = (1.2 \times 10^{-15} \text{ m})(294)^{1/3} = 8.0 \times 10^{-15} \text{ m} = 8.0 \text{ fm}.$$

11-7: For the electron, the magnetic potential energy is

$$U = \mu_B B = (5.788 \times 10^{-5} \text{ eV/T})(0.10 \text{ T}) = 5.8 \times 10^{-6} \text{ eV}.$$

For the proton, the magnetic potential energy is

$$U = \mu_p B = (2.793)(3.152 \times 10^{-8} \text{ eV/T})(0.10 \text{ T}) = 8.8 \times 10^{-9} \text{ eV}.$$

Nuclear Structure 87

11-9: (a) The ratio

$$\frac{\mu_p B}{kT} = \frac{(2.793)\,(3.152 \times 10^{-8} \text{ eV/T})\,(1.0 \text{ T})}{(8.617 \times 10^{-5} \text{ eV/K})\,(293 \text{ K})} = 3.49 \times 10^{-6}$$

is so small that the difference in populations of the two levels will be small. That is, each state can be assumed to have approximately $N/2$ protons, with the number of spin-up protons being $(N/2)e^{\mu_p B/kT}$ and the number of spin-down protons $(N/2)e^{-\mu_p B/kT}$. The difference is

$$\Delta N = N_- - N_+ = \frac{N}{2}\left(e^{\mu_p B/kT} - e^{-\mu_p B/kT}\right)$$

$$= N \sinh\left(\frac{\mu_p B}{kT}\right) = (10^6) \sinh(3.49 \times 10^{-6}) = 3.5.$$

In the above, the approximation $\sinh(x) \approx x$ is certainly valid. A more rigorous algebraic treatment, maintaining the same ratio of N_- to N_+ but requiring the sum to be exactly N is possible, leading to $\Delta N = N \tanh(\mu_p B/kT)$, but gives the same result.

(b) Repeating the above with $T = 20$ K gives $\Delta N = N \sinh(5.1 \times 10^{-5}) = 51$.

(c) Because the populations are so close, induced emission will nearly equal induced absorption, so there will be very little net absorption of the radiation.

(d) This is a two-level system, and could not be used as the basis for a laser.

11-11: The strong nuclear interaction, unlike the Coulombic or gravitational interactions, is short-range; the limited range limits the size of nuclei. (An explanation of why nuclear forces are short-range is given in Section 11.7 of the text.)

11-13: The nucleus $^{8}_{3}\text{Li}$ has three protons and five neutrons, and hence is an odd-odd nucleus, and is unstable, so $^{7}_{3}\text{Li}$ is the more stable of the two. The nucleus $^{15}_{6}\text{C}$ has three more neutrons (9) than protons (6); for a nucleus this small (in atomic number), that many excess neutrons do not serve to hold the nucleus together, and $^{13}_{6}\text{C}$ is more stable.

11-15: Using the values for the atomic masses and the constituent masses from the Appendix, the binding energy per nucleon of $^{20}_{10}\text{Ne}$ is

$$\frac{1}{20}\left[10\,(m_\text{H}) + 10\,(m_n) - m\binom{20}{10}\text{Ne}\right]$$

$$= \frac{1}{20}\left[10\,(1.007825 \text{ u}) + 10\,(1.008665 \text{ u}) - 19.992439 \text{ u}\right](931.49 \text{ MeV/u})$$

$$= 8.03 \text{ MeV}.$$

For $^{56}_{23}\text{Fe}$, the binding energy per nucleon is

$$\frac{1}{56}[26\,(1.007825\text{ u}) + 30\,(1.008665\text{ u}) - 55.934939\text{ u}]\,(931.49\text{ MeV/u}) = 8.79\text{ MeV}.$$

11-17: To remove a neutron from the ^4_2He nucleus, the energy needed is

$$m\left(^3_3\text{He}\right) + m_n - m\left(^4_2\text{He}\right)$$
$$= [3.016029\text{ u} + 1.008665\text{ u} - 4.002603\text{ u}]\,(931.49\text{ MeV/u}) = 20.58\text{ MeV}.$$

Then, to remove a proton, the energy needed is

$$m\left(^2_1\text{H}\right) + m_\text{H} - m\left(^3_2\text{He}\right)$$
$$= [2.014102\text{ u} + 1.007825\text{ u} - 3.016029\text{ u}]\,(931.49\text{ MeV/u}) = 5.49\text{ MeV}.$$

To separate the remaining proton and neutron, the energy needed is

$$m_n + m_\text{H} - m\left(^2_1\text{H}\right)$$
$$= [1.008665\text{ u} + 1.007825\text{ u} - 2.014102\text{ u}]\,(931.49\text{ MeV/u}) = 2.24\text{ MeV}.$$

The sum of these energies, to three significant figures, is 28.3 MeV.

The binding energy of ^4_2He is

$$2\,m_\text{H} + 2\,m_n - m\left(^4_2\text{He}\right)$$
$$= [2\,(1.007825\text{ u}) + 2\,(1.008665\text{ u}) - 4.002603\text{ u}]\,(931.49\text{ MeV/u}) = 28.3\text{ MeV},$$

the same as found above. Algebraically, the answers must be the same.

11-19: The electric potential energy of two protons separated by a distance 1.7 fm = 1.7×10^{-15} m is

$$\frac{e^2}{4\pi\epsilon_0\,r} = \frac{(8.988 \times 10^9\text{ N·m}^2/\text{C}^2)\,(1.602 \times 10^{-19}\text{ C})^2}{(1.7 \times 10^{-15}\text{ m})}$$
$$= 1.357 \times 10^{-13}\text{ J} = 0.85\text{ MeV}$$

to the given two significant figures.

The difference in binding energies is

$$\left[2\,m_\text{H} + m_n - m\left(^3_2\text{He}\right)\right] - \left[m_\text{H} + 2\,m_n - m\left(^3_1\text{H}\right)\right] = m_\text{H} - m_n - m\left(^3_2\text{He}\right) + m\left(^3_1\text{H}\right).$$

Using the atomic masses from the Appendix,

$$\Delta E = [1.007825\text{ u} - 1.008665\text{ u} - 3.016029\text{ u} + 3.016050\text{ u}]\,(931.49\text{ MeV/u})$$
$$= -0.763\text{ MeV},$$

Nuclear Structure

or -0.76 MeV to two significant figures, with the minus sign indicating that the tritium nucleus ^3_1H is more tightly bound than the ^3_2He nucleus. The magnitudes of the binding energy and the electric potential energy of the protons in ^3_2He are roughly the same, indicating that the most important contribution to the difference in binding energies is the mutual repulsion of the protons, an effect that is not present in ^3_1H. The closeness of magnitudes of the energies found is an indication that the nuclear forces must be very nearly independent of charge.

11-21: Using $A = 40$ and $Z = 20$ in Equation (11.18) (which makes the asymmetry term vanish) and the $+$ sign (even-even) for the pairing term, the predicted binding energy is

$$E_b = (14.1 \text{ MeV})(40) - (13.0 \text{ MeV})(40)^{2/3} - (0.595 \text{ MeV})\frac{(20)(19)}{(40)^{1/3}} + \frac{(33.5 \text{ MeV})}{(40)^{3/4}}$$

$$= 347.95 \text{ MeV}.$$

The actual binding energy is

$$20\, m_\text{H} + 20\, m_n - m\left(^{40}_{20}\text{Ca}\right)$$
$$= [20\,(1.007825 \text{ u}) + 20\,(1.008665 \text{ u}) - 39.962591 \text{ u}]\,(931.49 \text{ MeV/u})$$
$$= 342.05 \text{ MeV},$$

and the discrepancy is

$$\frac{347.95 \text{ MeV} - 342.05 \text{ MeV}}{342.05 \text{ MeV}} = 0.017 = 17\%.$$

11-23: (a) For mirror isobars of the form $^{2Z+1}_{Z}\text{X}$ and $^{2Z+1}_{Z+1}\text{Y}$, the difference in binding energy is (apart from a factor of c^2)

$$E_{Z+1} - E_Z = [(Z+1)\,m_\text{H} + Z\,m_n - M_{Z+1}] - [Z\,m_\text{H} + (Z+1)\,m_n - M_Z]$$
$$= -\Delta M - \Delta m,$$

where ΔM is the difference between the atomic masses of $^{2Z+1}_{Z}\text{X}$ and $^{2Z+1}_{Z+1}\text{Y}$, and $\Delta m = m_n - m_\text{H}$.

The difference between the coulomb energies is

$$\Delta E_c = \frac{3}{5}\frac{e^2}{4\pi\epsilon_0 R}[(Z+1)Z - Z(Z-1)] = \frac{3}{5}\frac{e^2}{4\pi\epsilon_0 R}\,2Z = \frac{3\,Z\,e^2}{10\pi\,\epsilon_0\,R}.$$

If this difference is equal to the *negative* of the difference in binding energies,

$$(\Delta M + \Delta m)c^2 = \frac{3}{10}\frac{Ze^2}{\pi\epsilon_0 R}.$$

Solving for R,

$$R = \frac{3}{10}\frac{Ze^2}{\pi\epsilon_0}\frac{1}{(\Delta M + \Delta m)c^2}.$$

(b) For the mirror isobars $^{15}_{7}$N and $^{15}_{8}$O, $Z = 7$ and

$(\Delta M + \Delta m)c^2$

$= [15.003065\ \text{u} - 15.000109\ \text{u} + 1.008665\ \text{u} - 1.007825\ \text{u}]\,(931.49\ \text{MeV/u})$

$= 3.536\ \text{MeV} = 5.665 \times 10^{-13}\ \text{J}.$

Using this in the expression for R found in part (a),

$$R = \frac{3}{10}\frac{(7)(1.602 \times 10^{-19}\ \text{C})^2}{\pi(8.854 \times 10^{-12}\ \text{C}^2/(\text{N} \cdot \text{m}^2))}\frac{1}{5.665 \times 10^{-13}\ \text{J}} = 3.42\ \text{fm}.$$

11-25: (a) Removing a neutron from an isotope of krypton leaves an isotope of krypton with mass number one less than that of the original isotope. For the given isotopes, the energy equivalents are

$m_n + m(^{80}_{36}\text{Kr}) - m(^{81}_{36}\text{Kr})$
$= [1.008665\ \text{u} + 79.916375\ \text{u} - 80.916578\ \text{u}]\,(931.49\ \text{MeV/u}) = 7.88\ \text{MeV}$

$m_n + m(^{81}_{36}\text{Kr}) - m(^{82}_{36}\text{Kr})$
$= [1.008665\ \text{u} + 80.916578\ \text{u} - 81.913483\ \text{u}]\,(931.49\ \text{MeV/u}) = 10.95\ \text{MeV}$

$m_n + m(^{82}_{36}\text{Kr}) - m(^{83}_{36}\text{Kr})$
$= [1.008665\ \text{u} + 81.913483\ \text{u} - 82.914134\ \text{u}]\,(931.49\ \text{MeV/u}) = 7.46\ \text{MeV}.$

(b) $^{82}_{36}$Kr has 36 protons and 46 neutrons, and so the neutrons are paired; the tendency of neutrons to pair together means removing a neutron from a $^{82}_{36}$Kr nucleus requires more energy.

11-27: In Equation (11.18), with $A = 127$ for each isobar, the coulomb energy term and the assymetry term will be different for the two nuclei. For $^{127}_{53}$I, $Z(Z-1) = (53)(52) = 2756$ and $(A-2Z)^2 = 441$. For $^{127}_{52}$Te, $Z(Z-1) = (52)(51) =$

Nuclear Structure 91

2652 and $(A-2Z)^2 = 529$. The difference in binding energies predicted by the liquid drop model is

$$\Delta E = E\left({}^{127}_{53}\text{I}\right) - \left({}^{127}_{52}\text{Te}\right) = -\frac{a_3}{A^{1/3}}(2756 - 2652) - \frac{a_4}{A}(361 - 529)$$

$$= -\frac{(0.595\text{ MeV})(104)}{(127)^{1/3}} - \frac{(19.0\text{ MeV})(-88)}{(127)}$$

$$= 0.855\text{ MeV},$$

and so ${}^{127}_{53}\text{I}$ is more stable, and ${}^{127}_{52}\text{Te}$ decays into ${}^{127}_{53}\text{I}$ by negative beta decay (electron emission).

11-29: A nucleon confined to a region of size $\Delta x = 2$ fm will have an uncertainty in momentum at least as large as $\dfrac{\hbar}{2\,\Delta x} = 2.63 \times 10^{-20}$ kg·m/s. The minimum kinetic energy a nucleon with this momentum would have is

$$\frac{(\Delta p)^2}{2\,m} = \frac{\left(2.63 \times 10^{-20}\text{ kg·m/s}\right)^2}{2\left(1.6736 \times 10^{-27}\text{ kg}\right)} = 2.1 \times 10^{-13}\text{ J} = 1.3\text{ MeV},$$

which is consistent with a potential well 35 MeV deep. Note that the nonrelativistic expression for kinetic energy is sufficient, and that the result is not changed if the mass of a neutron is used instead of the mass of a hydrogen atom.

Chapter 12 - Nuclear Transformations

In finding energy changes in nuclear transformations, accurate values of atomic masses are needed. The Appendix in the text, like most tables, gives the masses of *neutral* atoms; that is, the masses are the sum of the nuclear masses and the electrons, and include any binding energy of the electrons in their orbits. These electron binding energies are usually on the order of a few electronvolts and the precision of the masses given does not depend on the electron energies. As a specific example, the mass equivalent of the binding energy of an electron in the first Bohr orbit in a hydrogen atom is $(13.6 \text{ eV})/(931.49 \text{ MeV/u}) = 1.5 \times 10^{-8}$ u, a value too small to be reflected in the given atomic masses.

For four of the five possible nuclear transformation, it is sufficient to recognize that the initial and final states consist only of neutral atoms, and the electron masses need not be considered. For positron emission, the extra mass of the positron, which normally cannot exist in a netural atom, must be included.

Consider the reaction corresponding to positron emission (negative beta decay), represented as

$$ {}^{A}_{Z}X \longrightarrow {}^{A}_{Z-1}Y + e^+. $$

The change in the nuclear masses is related to the change in the neutral atomic masses by

$$ \left[m\left({}^{A}_{Z}X\right) - Z\, m_e \right] - \left[m\left({}^{A}_{Z-1}Y\right) - (Z-1)\, m_e \right] = \left[m\left({}^{A}_{Z}X\right) - m\left({}^{A}_{Z-1}Y\right) \right] + m_e, $$

where $m\left({}^{A}_{Z}X\right)$ and $m\left({}^{A}_{Z-1}Y\right)$ are the atomic masses as tabulated in the Appendix. However, the final state is *not* a neutral Y atom; there are still Z electrons and the emitted postitron. The atom is neutral, but it contains an extra electron-positron pair (for a short while), and this mass must be included in the final state. The result is that for positron emission, the energy released in the nuclear reaction is the energy of

$$ m\left({}^{A}_{Z}X\right) - m\left({}^{A}_{Z-1}Y\right) - 2\, m_e, $$

m_e being the common mass of the electron and positron.

The above discussion relates to Problems 12-27 and 12-62, among others.

Nuclear Transformations

12-1: 25 y is twice the half-life, so the fraction of the sample remaining is $2^{-2} = 1/4$.

12-3: The time of 1.00 s is small compared to the half-life of 37.3 min = 2232 s, and so N may be taken as a constant during this time. The probability that a particular nucleus will undergo beta decay is then the product of the decay constant and the time,

$$P = \lambda \Delta t = \frac{\ln 2}{T_{1/2}} \Delta t = \frac{\ln 2}{2232 \text{ s}} 1.00 \text{ s} = 3.1 \times 10^{-4}.$$

It should be noted that for calculators with sufficient precision, the above result is the same as

$$P = 1 - 2^{-\Delta t/T_{1/2}}.$$

12-5: The decay constant is $\lambda = \ln 2/T_{1/2}$; using this in Equation (12.5) and solving for the time t,

$$t = \frac{T_{1/2}}{\ln 2} \ln(N_0/N) = \frac{15.0 \text{ hr}}{\ln 2} \ln(5.0) = 34.8 \text{ hr}.$$

12-7: Using Equation (12.3), the half-life is related to the decay constant by

$$T_{1/2} = \frac{\ln 2}{\lambda} = \ln 2 \frac{N}{R}.$$

The number N of nuclei is the mass of the sample divided by the atomic mass. Approximating the atomic mass by the atomic number (see the Appendix),

$$T_{1/2} = \ln 2 \frac{(1.00 \times 10^{-3} \text{ kg})}{(226 \text{ u})(1.66054 \times 10^{-27} \text{ kg/u})} \frac{1}{(3.70 \times 10^{10} \text{ Bq})}$$
$$= 5.0 \times 10^{10} \text{ s} = 1.6 \times 10^3 \text{ y}.$$

12-9: From Equation (12.8), the activity is $R = \lambda N$ and the total number of atoms is the total mass divided by the mass of an atom, so

$$R = \lambda N = \frac{\ln 2}{T_{1/2}} \frac{m}{m\left(^{238}_{92}\text{U}\right)}$$
$$= \frac{\ln 2}{(4.5 \times 10^9 \text{ yr})(3.156 \times 10^7 \text{ s/y})} \frac{(1.0 \times 10^{-3} \text{ kg})}{(238 \text{ u})(1.66054 \times 10^{-27} \text{ kg/u})}$$
$$= 1.23 \text{ Bq},$$

keeping an extra significant figure.

12-11: The mass needed is the number of nuclei times the mass of each nucleus. Using Equation (12.8),

$$m = m(^{210}_{84}\text{Po})\frac{R}{\lambda} = m(^{210}_{84}\text{Po})\frac{RT_{1/2}}{\ln 2}$$

$$= (210\text{ u})(1.66054 \times 10^{-27}\text{ kg/u})\frac{(3.7 \times 10^8\text{ Bq})(138\text{ d})(86,400\text{ s/d})}{\ln 2}$$

$$= 2.22 \times 10^{-9}\text{ kg}.$$

12-13: Solving Equation (12.2) for the product λt gives $\lambda t = -\ln(R/R_0)$. Taking the time of the first measurement to be $t = 0$, so that the first measurement is taken as R_0, the natural logarithms of the activites as fractions of the initial activity are

$$\ln\left(\frac{80.5}{80.5}\right) = 0, \quad \ln\left(\frac{36.2}{80.5}\right) = -0.799, \quad \ln\left(\frac{16.3}{80.5}\right) = -1.597,$$

$$\ln\left(\frac{7.3}{80.5}\right) = -2.400, \quad \ln\left(\frac{3.3}{80.5}\right) = -3.194.$$

Four significant figures are not really warranted here, but are included for the intermediate calculations. Rounded to the hundredths place, the natural logarithms are $0, -0.80, -1.60, -2.40$ and -3.20, and so $\ln(R/R_0)$ is proportional to the time since the measurements began. From these data, it should be clear that to two significant figures, $-\lambda(1.0\text{ h}) = -0.80$. Using a plot to see the proportionality confirms this.

The plot for Problem 12-13 is shown on the next page.

The slope of the line that best fits the data is -0.80 h^{-1}, and so the experimental value for λ is 0.80 h^{-1} The plot was generated using a spreadsheet program that finds the slope of the best-fit line as -0.799 hr^{-1}, the same to two significant figures. The half-life $T_{1/2}$ is

$$T_{1/2} = \frac{\ln 2}{\lambda} = \frac{\ln 2}{(0.80\text{ h}^{-1})(1\text{ h}/60\text{ min})} = 52\text{ min}.$$

Plot for Problem 12-13

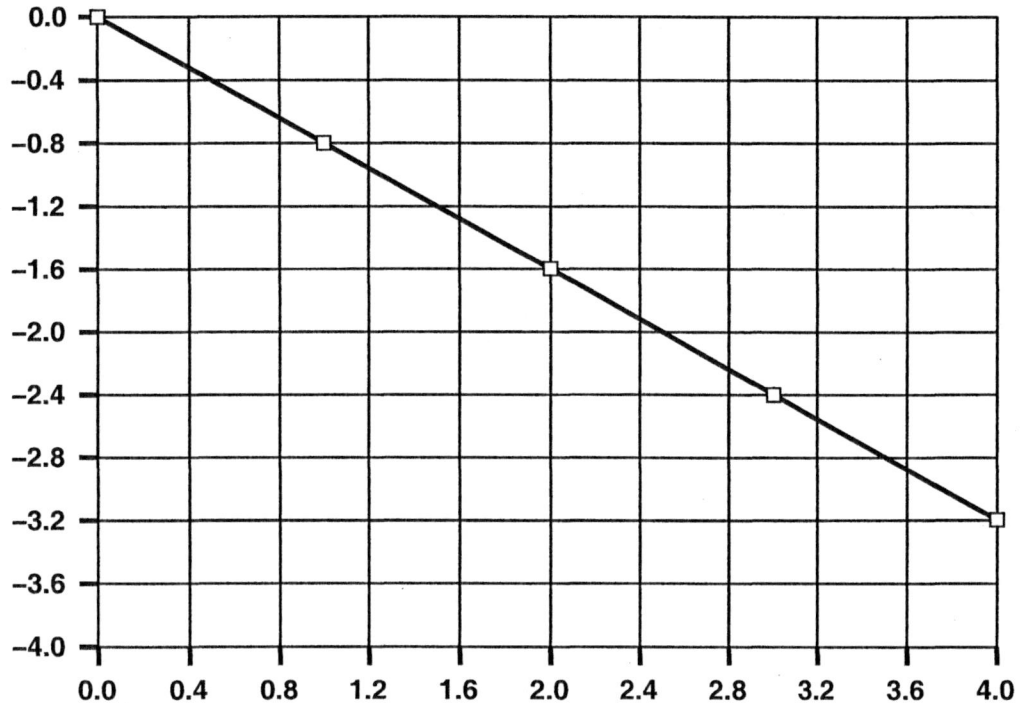

12-15: When the rock was formed, each nucleus that is currently a lead nucleus (^{206}Pb) was a uranium nulceus (^{238}U). The original mass M_0 of the sample would therefore have been

$$M_0 = M_U + M_{Pb}\frac{238}{206} = (4.00 \text{ mg})\left(1 + \left(\frac{1.00 \text{ mg}}{4.00 \text{ mg}}\right)\left(\frac{238}{206}\right)\right) = (4.00 \text{ mg})(1.289).$$

The ratio of the initial number of unranium nuclei to the current number is the same as the ratio of the masses. Solving Equation (12.5) for the time t,

$$t = \frac{\ln(N_0/N)}{\lambda} = \frac{\ln(M_0/M)}{(\ln 2/T_{1/2})} = \frac{T_{1/2}\ln(M_0/M)}{\ln 2}$$

$$= \frac{(4.47 \times 10^9 \text{ y})\ln(1.289)}{\ln 2} = 1.64 \times 10^9 \text{ y}.$$

12-17: See Example 12.5. The time since the wood was burned is

$$t = -\frac{1}{\lambda}\ln\left(\frac{R}{R_0}\right) = -\frac{T_{1/2}}{\ln 2}\ln\left(\frac{R}{R_0}\right) = -\frac{5760 \text{ y}}{\ln 2}\ln(0.18) = 1.4 \times 10^4 \text{ y}.$$

12-19: With the assumption that equal amounts of the nuclides (deonted A and B) were created, the ratio of the relative abundances as a function of time is

$$\frac{N_A}{N_B} = \frac{e^{-\lambda_A t}}{e^{-\lambda_B t}} = e^{(\lambda_B - \lambda_A)t}.$$

Solving for t,

$$t = \frac{\ln(N_A/N_B)}{\lambda_A - \lambda_B} = \frac{\ln(N_A/N_B)}{\ln 2 \left((1/T_{B,1/2}) - (1/T_{A,1/2})\right)}.$$

Using ^{238}U (the nuclide with the longer half-life, and hence the more abundant) for A and ^{235}U for B,

$$t = \frac{\ln(99.3/0.7)}{\ln 2 \left((7.0 \times 10^8 \text{ y})^{-1} - (4.5 \times 10^9 \text{ y})^{-1}\right)} = 5.9 \times 10^9 \text{ y}.$$

12-21: Each alpha decay lowers the mass number by 4, so the mass number of the lead isotope is $238 - 8\,(4) = 206$. Each alpha decay lowers the atomic number by 2, and each negative beta decay (electron emission) increases the atomic number by 1, so the atomic number of the lead isotope is $92 - 8\,(2) + 6 = 82$ (the isotope is given to be lead, and so the atomic number must be 82). The symbol is $^{206}_{82}$Pb. The energy released is the equivalent of

$$m(^{238}_{92}\text{U}) - m(^{206}_{82}\text{Pb}) - 8\,m(^{4}_{2}\text{He}) - 6\,m_e$$
$$= \left[238.050786 \text{ u} - 205.974455 \text{ u} - 8\,(4.002603 \text{ u}) - 6\,(5.84 \times 10^{-4} \text{ u})\right]$$
$$\times (931.49 \text{ MeV/u}) = 48.64 \text{ MeV}.$$

12-23: The kinetic energies of the alpha particle and the daughter nucleus are related by $\text{KE}_\alpha + \text{KE}_d = \mathcal{Q}$. The magnitude of the momenta of the alpha particle and the daughter nucleus must be the same, and from the nonrelativistic expression $p^2 = 2M\,\text{KE}$,

$$2\,M_\alpha\,\text{KE}_\alpha = 2\,M_d\,\text{KE}_d, \quad \text{or} \quad \frac{M_\alpha}{M_d}\,\text{KE}_\alpha = \text{KE}_d.$$

Substituting this expression into the expression for $\mathcal{Q}$,

$$\text{KE}_\alpha \left(1 + \frac{M_\alpha}{M_d}\right) = \text{KE}_\alpha \left(1 + \frac{4}{A-4}\right) = \text{KE}_\alpha \left(\frac{A}{A-4}\right), \quad \text{and}$$

$$\text{KE}_\alpha = \frac{A-4}{A}\,\mathcal{Q}.$$

Nuclear Transformations

12-25: An electron leaving a nucleus is attracted by the positive nuclear charge, which reduces the electron's energy. A positron leaving a nucleus is repelled by the nucleus and is accordingly accelerated outward, and so leaves the nucleus with greater energy.

12-27: See the remarks at the beginning of this chapter regarding positron emission.

The available energy when ^{7}Be decays to ^{7}Li is

$$m\left(^7_4\text{Be}\right) - m\left(^7_3\text{Li}\right) = 7.016930 \text{ u} - 7.016004 \text{ u} = 9.26 \times 10^{-4} \text{ u},$$

which is less than $2\, m_e = 1.0972$ u.

12-29: See Problem 12-26 and the discussion at the beginning of this chapter regarding proper treatment of the electron masses.

In electron emission, $^{80}_{35}\text{Br}$ becomes $^{80}_{36}\text{Kr}$. The difference between the masses of the neutral atoms is

$$79.918528 \text{ u} - 79.916375 \text{ u} = 2.153 \times 10^{-3} \text{ u} > 0,$$

so the reaction can occur. The energy released is

$$\left(2.153 \times 10^{-3} \text{ u}\right)(931.49 \text{ MeV/u}) = 2.01 \text{ MeV}.$$

In positron emission, $^{80}_{35}\text{Br}$ becomes $^{80}_{34}\text{Se}$. The atomic mass of the neutral copper atom must exceed the mass of the neutral selenium atom by twice the electron mass;

$$79.918528 \text{ u} - 79.916520 \text{ u} - 2\left(5.486 \times 10^{-4} \text{ u}\right) = 9.11 \times 10^{-4} \text{ u} > 0,$$

and so the reaction can occur. The energy released is

$$\left(9.11 \times 10^{-4} \text{ u}\right)(931.49 \text{ MeV/u}) = 0.85 \text{ MeV}.$$

In electron capture, $^{80}_{35}\text{Br}$ becomes $^{80}_{34}\text{Se}$, as in positron emission. The difference between the masses of the neutral atoms is

$$63.929766 \text{ u} - 63.927968 \text{ u} = 2.008 \times 10^{-3} \text{ u}.$$

The energy released is $\left(2.008 \times 10^{-3} \text{ u}\right)(931.49 \text{ MeV/u}) = 1.87$ MeV. Note that this is larger than the energy released in positron emission by twice the rest energy of an electron.

12-31: The minimum antineutrino energy needed is the energy equivalent of the difference between the rest masses of the final neutron and electron and the initial proton. Using the energy equivalents directly,

$$(m_n + m_e - m_p)\, c^2 = 939.57\text{ MeV} + 0.511\text{ MeV} - 938.28\text{ MeV} = 1.80\text{ MeV}.$$

12-33: The thirty-ninth proton in ^{89}Y is normally in a $p_{1/2}$ state and the next higher state available to this proton is a $g_{9/2}$ state; hence a radiative transition between the states has a low probability.

12-35: The neutron cross section decreases with increasing energy because the likelihood that a neutron will be captured depends on how much time the neutron spends near a particular nucleus; this time is inversely proportional to the neutron speed. The proton cross section is smaller at smaller energies because of the repulsive force exerted by the positive nuclear charge. See Problem 4-3 for a quantitative consideration.

12-37: The number density n is the ratio of the mass density and the mass of each atom,

$$n = \frac{\left(8.9 \times 10^3\text{ kg/m}^3\right)}{(59\text{ u})\left(1.66054 \times 10^{-27}\text{ kg/u}\right)} = 9.08 \times 10^{28}\text{ atoms/m}^3.$$

(a) The fraction that penetrates is given by Equation (12.20),

$$\frac{N}{N_0} = \exp(-n\,\sigma\, x)$$

$$= \exp\left(-\left(9.08 \times 10^{28}\text{ atoms/m}^3\right)\left(37 \times 10^{-28}\text{ m}^2\right)\left(1.0 \times 10^{-3}\text{ m}\right)\right)$$

$$= 0.71 = 71\%.$$

(b) From Equation (12.21), the mean free path is

$$\lambda = \frac{1}{n\,\sigma} = \frac{1}{\left(9.08 \times 10^{28}\text{ atoms/m}^3\right)\left(37 \times 10^{-28}\text{ m}^2\right)} = 3.0\text{ mm}.$$

12-39: Using $N = (1 - 0.99)\,N_0 = (0.01)\,N_0$ is Equation (12.20) and solving for x,

$$x = -\frac{\ln(0.01)}{n\,\sigma} = \frac{\ln(100)}{n\,\sigma}.$$

The number density n is the mass density divided by the mass per atom,

$$n = \frac{\left(2.2 \times 10^3\text{ kg/m}^3\right)}{(10\text{ u})\left(1.66054 \times 10^{-27}\text{ kg/u}\right)} = 1.32 \times 10^{29}\text{ m}^{-3}, \quad \text{and so}$$

$$x = \frac{\ln(100)}{\left(1.32 \times 10^{29}\text{ m}^{-3}\right)\left(4.0 \times 10^{-25}\text{ m}^2\right)} = 8.7 \times 10^{-5}\text{ m} = 0.087\text{ mm}.$$

Nuclear Transformations 99

12-41: In this situation the exposure time $\Delta t = 10.0$ h is much less than the half-life. This means that any decays of ^{60}Co that occur *while* the sample is exposed may be neglected (the correction is on the order of 2×10^{-3} %). The number N_{60} of ^{60}Co atoms after the sample has been exposed is the product of the number N_{59} of the original ^{59}Co atoms, the neutron flux S, the cross section σ and the exposure time Δt. The original number of ^{59}Co atoms is the ratio of the total mass to the mass of an atom. Combining,

$$\begin{aligned} N_{60} &= N_{59}\, S\, \sigma\, \Delta t \\ &= \frac{(10.0 \times 10^{-3}\text{ kg})}{(59\text{ u})(1.66054 \times 10^{-27}\text{ kg/u})} \left(5.00 \times 10^{17}\text{ neutrons/(m·s)}\right) \\ &\quad \times \left(37 \times 10^{-28}\text{ m}^2\right)\left(10.0\text{ h} \times 3600\,\frac{\text{s}}{\text{h}}\right) \\ &= 6.80 \times 10^{18}. \end{aligned}$$

The activity of the sample after exposure is

$$\begin{aligned} R = \lambda N_{60} &= \frac{\ln 2}{T_{1/2}} N_{60} \\ &= \frac{\ln 2}{(5.27\text{ y})(3.156 \times 10^7\text{ s/y})} \left(6.80 \times 10^{18}\right) \\ &= 2.83 \times 10^{10}\text{ Bq} = 0.77\text{ Ci}. \end{aligned}$$

Note that in the above, the quantity λ is a decay constant, not a mean free path.

12-43: The mass number of the unknown constituent must be $7 + 1 - 6 = 2$ and the atomic number must be $4 + 0 - 3 = 1$, and so the unknown nuclide is $^{2}_{1}\text{H}$ (a deuterium nucleus):

$$^{6}_{3}\text{Li} + {}^{2}_{1}\text{H} \longrightarrow {}^{7}_{4}\text{Be} + {}^{1}_{0}n.$$

The mass number of the unknown constituent must be $32 + 4 - 35 = 1$ and the atomic number must be $16 + 2 - 17 = 1$, and so the unknown nuclide is $^{1}_{1}\text{H}$ (a proton):

$$^{35}_{17}\text{Li} + {}^{1}_{1}\text{H} \longrightarrow {}^{32}_{16}\text{Be} + {}^{4}_{2}\text{He}.$$

The mass number of the unknown constituent must be $9 + 4 - 3(4) = 1$ and the atomic number must be $4 + 2 - 3(2) = 0$, and so the unknown nuclide is $^{1}_{0}n$ (a neutron):

$$^{9}_{4}\text{LBe} + {}^{4}_{2}\text{He} \longrightarrow 3\,{}^{4}_{2}\text{He} + {}^{1}_{0}n.$$

The mass number of the unknown constituent must be $79 + 2 - 2(1) = 79$ and the atomic number must be $35 + 1 = 36$, and so the unknown nuclide is $^{79}_{36}\text{Kr}$ (a rare isotope of krypton):

$$^{79}_{35}\text{Be} + ^{2}_{1}\text{H} \longrightarrow \, ^{79}_{36}\text{Kr} + 2\,^{1}_{0}n.$$

12-45: From Equation (12.26), with $m_A = m_p = 1$ u and $m_B = m_d = 2$ u,

$$\text{KE}_{\text{lab}} = \frac{3}{2} \text{KE}_{\text{cm}} = \frac{3}{2}(2.22 \text{ MeV}) = 3.33 \text{ MeV}.$$

12-47: Using Equation (12.24) with $m_A = 4$ u and $m_B = 16$ u, the speed of the center of mass of the system is (using a nonrelativistic calculation)

$$V = \frac{4}{20}v = \frac{1}{5}\sqrt{\frac{2\,\text{KE}_{\text{lab}}}{m}} = \frac{1}{5}\sqrt{\frac{2(5.0 \text{ MeV})}{(4 \text{ u})(931.49 \text{ MeV}/u\,c^2)}} = 0.014\,c = 3.1\times10^6 \text{ m/s}.$$

From Equation (12.16), the kinetic energy relative to the center of mass is

$$\text{KE}_{\text{cm}} = \frac{4}{5}\text{KE}_{\text{lab}} = 4 \text{ MeV}.$$

12-49: There are many possible ways to approach this problem; two are given here. Both methods must assume nonrelativistic motion.

Method (I): Let the original direction of the alpha particle be the x-direction, and let the plane of the interaction be the x-y plane. The original alpha particle has initial speed v_0 and final speed v'. The target nucleus has mass M and final speed V.

Conservation of momentum in both the x- and y-directions gives

$$m_\alpha v_0 = m_\alpha v' \cos 60° + M V \cos 30°$$
$$0 = m_\alpha v' \sin 60° - M V \sin 30°.$$

Multiplying the first equation by $\sin 30°$ and the second by $\cos 30°$ and adding,

$$m_\alpha v_0 \sin 30° = m_\alpha v' (\cos 60° \sin 30° + \sin 30° \cos 60°) = m_\alpha v' \sin 90° = m_\alpha v',$$

and so $v' = v_0 \sin 30° = v_0/2$.

Multiplying the first of the above equations by $\sin 60°$ and the second by $\cos 60°$ and subtracting,

$$m_\alpha v_0 \sin 60° = M V (\cos 30° \sin 60° + \sin 30° \cos°) = M V \sin 90° = M V,$$

and so $MV = m_\alpha v_0 \sin 60°$.

For an elastic collision, kinetic energy is conserved;

$$\frac{1}{2} m_\alpha v_0^2 = \frac{1}{2} m_\alpha v'^2 + \frac{1}{2} MV^2, \quad \text{or}$$

$$m_\alpha v_0^2 = m_\alpha (v_0 \sin 30°)^2 + \frac{(m_\alpha v_0 \sin 60°)^2}{M} = m_\alpha v_0^2 \frac{1}{4} + m_\alpha v_0^2 \frac{3}{4} \frac{m_\alpha}{M}.$$

This is solved for $M = m_\alpha$, and so the target has a mass number of 4.

In the above calculation, if the angle $60°$ is replaced by an arbitrary angle θ the result is

$$M = m_\alpha \frac{\sin^2 \theta}{1 - \sin^2(90° - \theta)} = m_\alpha \frac{\sin^2 \theta}{\cos^2(90° - \theta)} = m_\alpha,$$

suggesting perhaps an equivalent and more direct method of solution.

Method (II): Vector algebra may be used together with the fact that the particles move in perpendicular directions after the collision. Denote the initial momentum of the alpha particle by $\mathbf{p}_\alpha$, its final momentum by $\mathbf{p}'_\alpha$, and the final momentum of the target nucleus as $\mathbf{P}$. Then

$$\mathbf{p}_\alpha = \mathbf{p}'_\alpha + \mathbf{P}.$$

Taking the dot product of each side of the above equation with itself,

$$\mathbf{p}_\alpha \cdot \mathbf{p}_\alpha = (\mathbf{p}'_\alpha + \mathbf{P}) \cdot (\mathbf{p}'_\alpha + \mathbf{P}) = \mathbf{p}'_\alpha \cdot \mathbf{p}'_\alpha + \mathbf{P} \cdot \mathbf{P} + 2\mathbf{p}'_\alpha \cdot \mathbf{P}.$$

Because $\mathbf{p}'_\alpha$ and $\mathbf{P}$ are given as perpendicular, the last term on the right above vanishes, and so

$$p^2 = p'^2 + P^2.$$

For energy to be conserved,

$$\frac{p^2}{2 m_\alpha} = \frac{p'^2}{2 m_\alpha} + \frac{P^2}{2 M},$$

and comparison with the expression obtained from conservation of momentum gives the result $M = m_\alpha$.

12-51: (a) The excitation energy will be the kinetic energy of particle A in the center of mass frame, plus the Q value,

$$E^* = \left(\frac{m_B}{m_A + m_B}\right) \text{KE}_A + Q = \left(\frac{m_C - m_A}{m_C}\right) \text{KE}_A + Q = \left(1 - \frac{m_A}{m_C}\right) \text{KE}_A + Q.$$

In this expression, the approximation $m_c \approx m_A + m_B$, valid when $Q \ll m_C c^2$, has been made.

(b) The Q value for this reaction is

$$[m(^{15}_7\text{N}) + m_p - m(^{16}_8\text{O})]\,c^2$$
$$= [15.000109 \text{ u} + 1.007825 \text{ u} - 15.994915 \text{ u}]\,(931.49 \text{ MeV/u})$$
$$= 12.13 \text{ MeV}.$$

Solving the above expression for KE_A,

$$\text{KE}_A = (E^* - Q)\frac{m_C}{m_C - m_A} = (16.2 \text{ MeV} - 12.13 \text{ MeV})\frac{16}{15} = 4.43 \text{ MeV}.$$

12-53: The neutron to proton ratio required for stability decreeases with decreasing mass number A, hence there is an excess of neutrons when fission occurs. Some of the excess neutrons are released directly, and the others change to protons by beta decay in the fission fragments.

12-55: Using Equation (11.1) for the nuclear radii, the centers of the nuclei are separated by $R_0\left(A_1^{1/3} + A_2^{1/3}\right)$, and the electrostatic potential energy is

$$U = \frac{Q_1 Q_2}{4\pi \epsilon_0 R_0 \left(A_1^{1/3} + A_2^{1/3}\right)}$$

$$= \left(8.988 \times 10^9 \text{ N·m}^2/\text{C}^2\right) \frac{(38)(54)\left(1.602 \times 10^{-19} \text{ C}\right)^2}{(1.2 \times 10^{-15} \text{ m})\left(94^{1/3} + 140^{1/3}\right)}$$

$$= 4.05 \times 10^{-11} \text{ J} = 253 \text{ MeV}.$$

12-57: The ^1_1H nuclei in ordinary water are protons, which readily capture neutrons to form ^2_1H (deuterium) nuclei. The neutrons cannot contribute to the chain recation in a reactor, so a reactor using ordinary water as a moderator needs enriched unranium with a greater content of the fissionable ^{235}U isotope to function. Deuterium nuclei are less likely to capture neutrons than are protons; hence a reactor moderated with heavy water can operate with ordinary uranium as fuel.

Nuclear Transformations

12-59: Let the initial speed of the particle with mass m_1 be v_1 and the final speeds by v_1' and v_2'.

(a) Conservation of momentum gives

$$m_1 v_1 + m_2 v_2 = m_1 v_1' + m_2 v_2' \qquad \text{or} \qquad m_1(v_1 - v_1') = m_2 v_2'$$

and conservation of kinetic energy gives

$$\frac{1}{2} m_1 v_1^2 = \frac{1}{2} m_1 v_1'^2 + \frac{1}{2} m_2 v_2'^2 \qquad \text{or} \qquad m_1\left(v_1^2 - v_1'^2\right) = m_2 v_2'^2.$$

If $v_1' = v_1$, there is no collision. If there is a collision, dividing the equation obtained from conservation of kinetic energy by the equation obtained from conservation of momentum, and using $v_1^2 - v_1'^2 = (v_1 - v_1')(v_1 + v_1')$ gives

$$v_1 + v_1' = v_2', \qquad \text{or} \qquad v_1 = v_2' - v_1'.$$

This standard result from classical mechanics is often interpreted as the relative speeds of the particles being the same before and after the collision, a result that holds even if the second particle is moving initially; that is, the relative speed is independent of the frame of the observer if the observer and the particles are not moving relativistically.

The two equations

$$m_1 v_1 = m_1 v_1' + m_2 v_2' \qquad \text{and} \qquad v_1 = v_2' - v_1'$$

are solved for

$$v_2' = \frac{2 m_1}{m_1 + m_2} v_1.$$

The desired ratio of kinetic energies is

$$\frac{\text{KE}_2'}{\text{KE}_1} = \frac{(1/2) m_2 v_2'^2}{(1/2) m_1 v_1^2} = \frac{4 m_1 m_2}{(m_1 + m_2)^2} = 4 \frac{(m_2/m_1)}{(1 + (m_2/m_1))^2}.$$

(b) Virtually all of the neutron's kinetic energy will be transferred to the proton, as the masses are almost indentical (use of the actual masses gives $1 - 2 \times 10^{-7}$). For a collision with a deuteron, the ratio m_2/m_1 is essentially 2, and $4(2)/(3)^2 = 0.89 = 89\%$. For a collision with a ^{12}C nucleus, $m_2/m_1 = 12$ and $4(12)/(13)^2 = 0.28 = 28\%$. For a collision with a ^{238}U nucleus, $m_2/m_1 = 238$, and $4(238)/(239)^2 = 0.017 = 1.7\%$.

12-61: The minimum kinetic energy the proton must have is the electrostatic potential energy of the proton-nucleus combination when the proton is at the nuclear surface. Using Equation (1.11) to give the radius of the nucleus and using R_0 for the radius of the proton,

$$U = \frac{Z e^2}{4\pi \epsilon_0 R_0 \left(1 + A^{1/3}\right)} = \left(8.988 \times 10^9 \text{ N·m}^2/\text{C}^2\right) \frac{(6) \left(1.602 \times 10^{-19} \text{ C}\right)^2}{\left(1.2 \times 10^{-15} \text{ m}\right)\left(1 + 12^{1/3}\right)}$$

$$= 3.51 \times 10^{-10} \text{ J} = 2.19 \text{ MeV}.$$

12-63: (a) The electrostatic energy of the deuterons separated by the given distance is

$$U = \frac{e^2}{4\pi \epsilon_0 r} = \left(8.988 \times 10^9 \text{ N·m}^2/\text{C}^2\right) \frac{\left(1.602 \times 10^{-19} \text{ C}\right)^2}{\left(5 \times 10^{-15} \text{ m}\right)}$$

$$= 4.6 \times 10^{-14} \text{ J} = 2.9 \times 10^5 \text{ eV}.$$

For the average translational kinetic energy $(3/2)kT = U$,

$$T = \frac{2}{3} \frac{U}{k} = \frac{2}{3} \frac{2.9 \times 10^5 \text{ eV}}{8.617 \times 10^{-5} \text{ eV/K}} = 2.2 \times 10^9 \text{ K}.$$

(b) This temperature corresponds to the average deuteron energy, but many deuterons have considerably higher energies than the average. Also, quantum-mechanical tunneling through the potential barrier can occur, permitting deuterons to react despite having insufficient energy to come together classically.

Chapter 13 - Elementary Particles

13-1: (a) From the uncertainty principle, with $\Delta E = 2\,mc^2$, the uncertainty in the time that such an electron-positron pair may exist is

$$\Delta t \geq \frac{\hbar/2}{2\,mc^2} = \frac{(6.582 \times 10^{-16} \text{ eV·s})}{4\,(0.511 \times 10^6 \text{ eV})} = 3.22 \times 10^{-22} \text{ s}.$$

(b) The strong electric field of the nucleus separates the electron and positron sufficiently so that they cannot recombine afterward to reconstitute the photon.

13-3: A relativistic calculation, including the recoil of the Λ^0 particle, is necessary. The Λ^0 particle and the photon will have momentum of magnitude

$$p_{\Lambda^0} = p_\gamma = \frac{E_\gamma}{c}.$$

This may be used to relate the total energy of the Λ^0 particle to the photon energy,

$$p_{\Lambda^0}^2 c^2 = E_{\Lambda^0}^2 - \left(m_{\Lambda^0} c^2\right)^2 = E_\gamma^2.$$

From conservation of energy,

$$m_{\Sigma^0} c^2 = E_{\Lambda^0} + E_\gamma, \quad \text{and}$$

$$E_{\Lambda^0}^2 = \left(m_{\Sigma^0} c^2\right)^2 + E_\gamma^2 - 2\,m_{\Sigma^0} c^2\, E_\gamma.$$

Equating the two expressions for $E_{\Lambda^0}^2$, canceling the E_γ^2 term and solving for the photon energy,

$$E_\gamma = \frac{\left(m_{\Sigma^0} c^2\right)^2 - \left(m_{\Lambda^0} c^2\right)^2}{2\,m_{\Sigma^0} c^2} = \frac{(1193 \text{ MeV})^2 - (1116 \text{ MeV})^2}{2\,(1193 \text{ MeV})} = 74.5 \text{ MeV}.$$

The above expression for the photon energy may be expressed as

$$E_\gamma = \left(m_{\Sigma^0} c^2 - m_{\Lambda^0} c^2\right)\left(1 - \frac{m_{\Sigma^0} - m_{\Lambda^0}}{2\,m_{\Sigma^0}}\right),$$

which shows that in the nonrelativistic limit, $m_{\Sigma^0} \gg m_{\Sigma^0} - m_{\Lambda^0}$, the photon energy is just the difference between the rest mass energies of the particles.

13-5: The minimum photon energy would be for the situation where all three of the final electrons have the same momentum (no relative motion, and hence no motion in the center of mass frame). Denote this common momentum magnitude by p'. Assuming the initial electron to be at rest, the initial photon momentum would be $3p'$, and the initial photon energy is $E_\gamma = 3p'c$. From conservation of energy,

$$E_\gamma + m_e c^2 = 3E', \qquad \left(E_\gamma + m_e c^2\right)^2 = 9 E'^2 = 9\left((p'c)^2 + \left(m_e c^2\right)^2\right),$$

where the common final energy of each electron is E'. Squaring the binomial and using $9(p'c)^2 = E_\gamma^2$,

$$E_\gamma^2 + 2 E_\gamma m_e c^2 + \left(m_e c^2\right)^2 = E_\gamma^2 + 9\left(m_e c^2\right)^2$$
$$2 E_\gamma m_e c^2 = 8 \left(m_e c^2\right)^2.$$

from which $E_\gamma = 4\left(m_e c^2\right)^2$ follows.

As an equivalent alternative, consider the center of mass frame (more accurately, the center of momentum frame) in which the three electrons are created at rest, and hence with zero momentum. In this frame, the initial momentum of the photon and electron would have the same magnitude p_0, and the photon would have energy $E_0 = p_0 c$. From conservation of energy,

$$E_0 + \sqrt{m_e^2 c^4 + p_0^2 c^2} = 3 m_e c^2, \qquad \sqrt{m_e^2 c^4 + E_0^2} = 3 m_e c^2 - E_0.$$

Squaring both sides of the second relation and canceling E_0^2 gives

$$m_e^2 c^4 = 9 m_e^2 c^4 - 6 m_e c^2 E_0, \qquad \text{and} \qquad E_0 = \frac{4}{3} m_e c^2, \qquad p_0 = \frac{4}{3} m_e c.$$

From Equation (1.16), then,

$$\frac{(v/c)}{\sqrt{1-(v/c)^2}} = \frac{4}{3},$$

which is solved for $(v/c) = (4/5)$, and this must be the speed of the center of mass frame relative to the frame where the electron was initially at rest (the lab frame). From Equation (1.8), the energy of the photon in the lab frame would be higher than E_0,

$$E_\gamma = E_0 \sqrt{\frac{1+(v/c)}{1-(v/c)}} = \frac{4}{3} m_e c^2 \sqrt{(9/5)/(1/5)} = 4 m_e c^2.$$

Elementary Particles 107

13-7: Denote the initial pion momentum by p_π and the gamma-ray momentum magnitude by p_γ. From conservation of momentum, $p_\pi = 2 p_\gamma \cos\theta$, where θ is the half-angle between the gamma-ray paths (θ is the angle that each gamma ray makes with respect to the initial direction of the pion's momentum). The initial energy of the pion in terms of its momentum is

$$\sqrt{p_\pi^2 c^2 + m_\pi^2 c^4} = 2 m_\pi c^2, \qquad \text{so} \qquad p_\pi = \sqrt{3}\, m_\pi c.$$

From conservation of energy,

$$\text{KE}_\pi + m_\pi c^2 = w m_\pi c^2 = 2 p_\gamma c, \qquad \text{so} \qquad m_\pi c = p_\gamma \qquad \text{and} \qquad p_\pi = \sqrt{3}\, p_\gamma.$$

Equating the two expressions relating p_π and p_γ yields $\cos\theta = \dfrac{\sqrt{3}}{2}$, and so $\theta = 30°$ (giving θ in radians as $\pi/6$ might cause confusion). The angle between the two gamma rays is $2\theta = 60°$.

13-9: (a) does not conserve baryon number. (b) can occur. (c) does not conserve charge. (d) can occur.

13-11: The spontaneous appearance of neutron-antineutron pairs would violate conservation of energy.

13-13: The other particle must have charge $Q = 0$ and muonic lepton number $L_\mu = +1$. The only such particle is ν_μ, a μ-neutrino.

13-15: The other particle must have charge $Q = -e$, baryon number $B = +1$ and strangeness $S = -2$. (The original negative kaon had strangeness $S = +1$ and the final kaon has strangeness $S = -1$.) From Table 13.1, the only such particle is the Ξ^-, a negative xi particle.

13-17: Quarks are fermions, and if quarks with the same color combined to form a hadron, the exclusion principle would be violated. If the spins of quarks were integral instead of half-integral, the exclusion principle would not apply, and quarks of the same color could be the constituents of a hadron.

13-19: From Table 13.4, the sum of the charges of two u quarks and an s quark is $2\frac{2}{3}e - \frac{1}{3}e = +e$, and the particle is Σ^+ as given in Table 13.3

13-21: The combination uus has strangeness $S = -1$ and charge $Q = 2\frac{2}{3}e - \frac{1}{3}e = +e$, and from Table 13.3 the particle is Σ^+.

13-23: Only the strong interaction, which affects only hadrons, can produce such rapid decays.

13-25: Because a positron and a neutrino are emitted, the weak interaction is involved; the weak interaction is so much feebler than the strong interaction that the initial reaction of the proton-proton cycle has a low probability of occuring even when the protons are energetic enough to overcome the Coulomb barrier.

13-27: (a) If the angular separation of two spots is the angle θ, in radians, then $s = r\theta$ and

$$\frac{ds}{dt} = \frac{dr}{dt}\theta = \frac{dr}{dt}\frac{s}{r} = \frac{1}{r}\frac{dr}{dt}s.$$

where the constancy of the angular separation is used to set $\frac{d\theta}{dt} = 0$. The radius r and the rate of change $\frac{dr}{dt}$ are the same for all points on the sphere at any time.

(b) The parameter H is then the factor multiplying s in the above expression for $\frac{ds}{dt}$;

$$H = \frac{1}{r}\frac{dr}{dt},$$

which is sometimes expressed as

$$H = \frac{d}{dt}\ln(r).$$

In this form, it is readily seen that if $\ln r = \ln r_0 + kt$, H is constant and equal to k, but if not, H is not constant. Thus, H will be constant if $r = r_0 e^{kt}$. If this is the case with $k > 0$, the balloon is expanding at an every-increasing rate. This phenomenon would be like the proposed "inflationary universe" theory of the early universe. If $k < 0$, the balloon is shrinking, gradually approaching zero radius. If $k = 0$, $H = 0$ and the balloon is neither expanding or shrinking.

Notes

Notes

Notes

Notes

Notes

Notes

Notes

Notes

Notes

Notes

Notes